Carbon Management Technology Conference (CMTC 2019)

AA003002

Houston, Texas, USA
15-18 July 2019

ISBN: 978-1-7138-0091-0

Printed from e-media with permission by:

Curran Associates, Inc.
57 Morehouse Lane
Red Hook, NY 12571

Some format issues inherent in the e-media version may also appear in this print version.

Printed with permission by Curran Associates, Inc. (2020)

For permission requests, please contact AIChE
at the address below.

AIChE
120 Wall Street, FL 23
New York, NY 10005-4020

Phone: (800) 242-4363
Fax: (203) 775-5177

www.aiche.org

Additional copies of this publication are available from:

Curran Associates, Inc.
57 Morehouse Lane
Red Hook, NY 12571 USA
Phone: 845-758-0400
Fax: 845-758-2633
Email: curran@proceedings.com
Web: www.proceedings.com

Carbon Management Technology Conference 2019 (CMTC 2019) proceedings

General Submissions

Sessions:

Session 8: Storage & EOR (III)

Papers:

Session 9: Economic Analysis and Modeling & Business and Financial Approaches

Papers:

Session 10: Carbon Capture - Transformational and 2nd Generation (III) & Monitoring and Characterization Protocols and Technologies

Papers:

Session 11: Laboratory Investigations

Papers:

Session 12: Policy, Legislation, and Permitting Approaches & Country and Regional Climate Change

Papers:

Session 15: Carbon Utilization (II)

Papers:

Session 16: Industry and Energy Sector CCUS Projects (II)

Papers:

Poster Session

Papers:

90 Potential Application of CCS Technology for Reducing the Carbon Footprint Created during Deepsea Mining Operations

Shukai Wu

91 Rapid Evaluation Model of Enhanced Oil Recovery of CO2 Flooding Based on Modification Fractional Flow Theory

Qiang Yu
Hongli Xi
Xiaoliang Zhao

92 Reaction Experiments of CO2-Rock-Water-Crude Oil and Its Effect on CO2 EOR

Qiang Yu
Xiaoliang Zhao
Hongli Xi

93 Refiner's Industrial Scale Facility Uses CO2 in Stack Gas to Enhance Crop Production

Brian Kolodji

94 Synthesis Metal-Doped Mesoporous Resin for Carbon Dioxide Capturing and Electrochemical Conversion

Mahboubeh Nabavinia

95 The Effect of Asphaltene Structure on Wettability Alteration during Enhanced Oil Recovery Using Carbonated Water

Mohammad Alqam
Nadrah Alawani

96 Understanding the Subsurface Structural Formation for Potential CO2 Storage and Capture in Geologic Formation

Uchenna Nwoke

Poster Session (Registered)

Papers:

97 3D-Printed Monoliths for CO2 Capture and Utilization Applications

Fateme Rezaei

98 A Review of the Mechanisms of Miscible and Immiscible Carbon Dioxide Injection in Shale Reservoirs for Storage and Increased Oil Recovery Applications

Sherif Fakher
Abdulmohsin Imqam

99 An Experimental Study of the Factors Impacting Asphaltene Stability in Crude Oil during Carbon Dioxide Injection in Nanopores

Sherif Fakher

Bench Scale Testing of Next Generation Hollow Fiber Membrane Modules

Authors:
Fu, S., American Air Liquide D.I.C
Hasse, D., Air Liquide D.I.C
Hamilton, A., Air Liquide
Augustine, A., Air Liquide
Kulkarni, S., Air Liquide D.I.C

Air Liquide is developing a next generation polyimide membrane material (PI-2) for application in our cold membrane hybrid process for CO_2 capture. Combining the PI-2 membrane operation with an integrated CO_2 compression and purification unit significantly reduces the overall cost of CO_2 capture. The current work is aimed at developing the PI-2 material to a commercial scale, then validating its performance under real flue gas conditions.

Previous work showed promising results for the cold membrane hybrid process utilizing the commercially available Air Liquide PI-1 membrane bundles. The cold membrane technology was demonstrated at the National Carbon Capture Center (NCCC) with a 0.3 MWe field-test unit incorporating 12â€ diameter PI-1 commercial bundle. The cost of 90% CO_2 capture from a 550 MWe net coal power plant was estimated to be between 40 and $45/tonne. The novel PI-2 membrane was tested at a smaller scale (1â€ modules) and exhibited up to 10 times the CO_2permeance and similar CO_2/N_2 selectivity as the PI-1 membrane. This corresponds to a potential cost reduction of $3/tonne for CO_2 capture.

The current project seeks to develop this next generation PI-2 material to a technology readiness level of 5 via manufacturing scale-up and testing with 600 - 1,000 Nm^3/h of real flue gas, 0.2 - 0.3 MWe equivalents. 6â€ diameter commercial membrane bundles have been fabricated and tested at NCCC. These bundles met the project success criteria with a productivity of over 640 Nm^3/hr and permeate purity in excess of 60%, at 90% CO_2 recovery. An advanced bundle formation technique, which further lowers the bundle cost and CO_2 capture cost, is also being developed under the scope of this project. Further manufacturing scale-up is underway to verify the technique. A final techno-economic analysis (TEA) for a 550 MWe net power plant has been initiated.

Doe's Efforts to Accelerate Transformational Carbon Capture R&D through Advanced Computing and Manufacturing

Authors:
Litynski, J., U. S. Department of Energy
Brickett, L., U.S. DOE National Energy Technology Laboratory
Matuszewski, M. S., AristoSys, LLC, Contractor to National Energy Technology Laboratory

The steel industry is among the most energy-intensive industries of manufacturing sector emitting significant volumes of carbon dioxide (CO2). On average, 1.9 tonnes of CO2 are emitted for every tonne of steel produced [1] . According to the International Energy Agency, the iron and steel industry accounts for approximately 6.7% of total world CO2 emissions.[1] The industries sector accounted for 21.7 percent of the total GHG emissions or 412.55 million tons of CO2 eq in the year 2005. In the year 2007, iron & steel production resulted in 6.2 percent emissions (117.32 million tons).[2] India have committed 33-35 per cent cut in carbon emissions by 2030 in from 2005 level in the target submitted to the United Nations Framework Convention (UNFCCC) on Climate Change for a global climate pact.[3] The paper discusses various carbon capture storage (CCS) and carbon capture utilization (CCU) technologies are being developed and implemented in pilot plants / commercial plants in Europe and USA ,which may be implemented on commercial scale in Indian steel industries. The Indian steel sector, due to its scale of production and energy intensive nature, represents a sizeable opportunity to reduce CO_2 emissions. At 2018 energy and emission intensities, this production would entail 200 $MTCO_2$-eq as emissions

Keywords: iron and steel, production, energy intensity, CO2 emissions, CO2 mitigation, CCS & CCU

Evaluated Results for Pilot Testing of CO2 Capture By Piperazine Regeneratec By the Advanced Flash Stripper

Authors:
Chen, E., Gary T Rochelle
Rochelle, G. T., The University of Texas at Austin
Gao, T., University of Texas at Austin
Wu, Y., The University of Texas at Austin
Akinpelumi, K., The University of Texas at Austin
Fischer, K., The University of Texas at Austin
Liu, C. T., The University of Texas at Austin
Farmer, K., AECOM process technologies
Sexton, A., Trimeric Corporation

CO_2 capture from coal-fired flue gas was demonstrated with 5m piperazine (30 wt %) regenerated by the Advanced Flash Stripper in 2018 for more than 2000 hours in a 26-inch absorber at the National Carbon Capture Center. 90 to 99% CO_2 removal was achieved with a heat duty of 2.1 to 2.3 GJ/tonne CO_2 removed. The stripper was operating at 6 bar with a sump temperature of 150°C. The absorber used two 6-meter beds of structured packing with solvent intercooling between the beds. Amine oxidation was less than 0.2 kg/tonne CO_2. Amine aerosol emissions were managed to be less than 1 ppm with 2 ppm of SO_3 by operating with a warm lean feed and with a two-stage water wash. Carbon steel was frequently protected by a $FeCO_3$ film, both carbon steel and stainless steel experienced unacceptable corrosion at the hot rich condition at 150°C.

Reducing CO2 Capture Cost By 30% Using Advanced KM CDR ProcessTM

Authors:
Kamijo, T., Mitsubishi Heavy Industries, Ltd.

Upon completion of the worldâ€™s largest post-combustion CO_2 capture plant â€“ the Petra Nova Project â€“ Mitsubishi Heavy Industries engineering (MHIEng) has demonstrated that commercial-scale CO_2 capture for coal-fired power plant is technically feasible. Applying lessons learned from not only Petra Nova project but also other commercial plants, MHIEng has developed the Advanced KM CDR Process™, providing superior performance with significantly lower capital cost. The cost reduction is a result of smaller flue gas quencher and CO_2 absorber designs, reduced design redundancy, and modular design. These modifications can reduce the total EPC cost of the next CO_2 capture and compression plant by nearly 30% compared to MHIEngâ€™s conventional process. In addition, MHIEng is currently developing a new solvent (KS-21) which will further improve overall plant economics. KS-21 solvent is expected to be superior to MHIEngâ€™s existing KS-1â„¢ solvent in stability and lower volatility. The pilot plant test shows that KS-21 has 50% lower amine emissions than KS-1â„¢ with comparable energy performance. With better thermal stability, it is expected that the new solvent can be operated at elevated regenerator temperature and pressure and thus reduce downstream CO_2 compression work.

Transformational Graphene Oxide-Based Membrane Process for Post-Combustion CO2 Capture

Authors:
Li, S., GTI
Yu, M., Rensselaer Polytechnic Institute

The Department of Energy (DOE), through its Carbon Capture Program, is investigating transformational, low-cost CO_2 separation and capture technologies that can be installed in new or retrofitted into existing pulverized coal (PC) power plants for CO_2 capture with a cost of electricity at least 30% lower than a supercritical PC with CO_2 capture, or approximately $30 per tonne of CO_2 captured.

Supported by DOE, we are developing a transformational graphene oxide (GO)-based membrane process that can be installed in new or retrofitted into existing pulverized coal (PC) or natural gas power plants for CO_2 capture with 95% CO_2 purity. The GO-based membrane technology builds on our pioneering work on the ultrathin, GO-based membranes published in *Science* (Li et al., *Science*, Vol. 342, no. 6154 (2013) 95) and *Nature Communications* (Zhou et al., *Nature Communications*, Vol. *8,* doi:10.1038).

A facile and scalable coating process was developed to prepare ultrathin graphene oxide (GO)-based membranes on polyethersulfone hollow fibers. The GO-based membranes developed to date exhibited extraordinary separation performance under simulated flue gas conditions with CO_2 permeance of 1,020 GPU and a CO_2/N_2 selectivity as high as 680. The selectivity is much higher than state-of-the-art membranes. In addition, after extensive material and structure optimization, we have also developed high-flux membranes with CO_2 permeance as high as 2,500 GPU using GO quantum dots as a membrane building block.

Our transformational GO-based membrane process integrates the high-selectivity and high-flux GO-based membranes and offers a new opportunity to explore further reductions in the cost of CO_2 capture. Our process, functionally, has the benefits of a hybrid system but the simplicity of a membrane system which also reduces the up-front installation costs and footprint and does not add circulating liquids to the power plant environment. Therefore, it facilitates the ease of integration into a power plant.

Clarifying the Relationship of Enhanced Oil Recovery with Associated Storage of CO2

Authors:
Hamling, J. A., University of North Dakota
Wildgust, N., University of North Dakota
Sorensen, J. A., University of North Dakota
Peck, W. D., University of North Dakota
Gorecki, C. D., University of North Dakota

Carbon dioxide (CO_2) enhanced oil recovery (CO_2 EOR) is a proven technology to increase production from, and extend the lifespan of, conventional oil fields. The financial benefit of incremental oil production provides the motivation for CO_2EOR operations. Growing interest in capturing anthropogenic CO_2 across a range of industrial sectors, motivated by recent 45Q tax credits, could translate to significant opportunities to expand the scale of CO_2 EOR operations across the United States. When anthropogenic CO_2 sources are used, widespread misconceptions still exist concerning the relationship between the CO_2 EOR process and the associated CO_2 storage that occurs as a consequence.

Metrics used by reservoir engineers, including CO_2 utilization factors, CO_2retention, and incremental oil recovery, allow determination of EOR efficiency and the associated storage. The ability of geologic reservoirs to retain buoyant hydrocarbon fluids over geological timescales, combined with a CO_2 recycling system, maximizes the use of purchased CO_2 and ensures virtually all purchased CO_2 is ultimately stored.

Over 6 Mt of associated CO_2 storage incidental to CO_2 EOR has occurred at the Bell Creek oil field, illustrating the relationship between associated storage and EOR. Injection and production data collected from Bell Creek illustrate temporal relationships between purchased, recycled, and injected CO_2 quantities, utilization rates, and associated storage. A research-monitoring program conducted at Bell Creek by the Plains CO_2 Reduction (PCOR) Partnership effectively tracked injected CO_2 in the reservoir and demonstrated an absence of discernible impacts to shallow groundwater, surface water, and other environmental receptors.

The Energy & Environmental Research Center (EERC) has used this detailed knowledge of the generic relationship between EOR and associated storage to determine the potential economic and carbon management benefits of CO_2 EOR deployment in conventional oil fields across North Dakota.

Evaluating Foam Stability Using Tailored Water Chemistry for CO2 Mobility Control Applications

Authors:
Al Yousef, Z., Saudi Aramco
Ayirala, N., Saudi Aramco
AlAbdulwahab, A., Saudi Aramco
Kokal, S., Saudi Aramco

Generating in-situ foam with CO_2 is regarded as one of the most promising techniques to overcome gas mobility issues and improve sweep efficiency in EOR processes. Gravity override, viscous fingering and channeling through permeable zones are the major limiting factors that can impair the efficiency of gas floods. Generating strong and stable foam while injecting CO_2 is one way to achieve in-depth conformance improvement in the reservoir.

The foams are generated by mixing the injection gas with injection water containing surfactant in the laboratory. A tailored water chemistry (formulated low salinity water) has been evaluated in comparison to using typical high salinity injection water in surfactant solution to determine its overall effect on the produced foam. Using foam rheology apparatus, the foam stabilization factors were analyzed and quantified by measuring the foam-life and the bubble sizes of different surfactants in varying salinity water solutions.

The results clearly demonstrated that the use of tailored water chemistry improved the stability of produced foams when compared to high salinity water. Low salinity tailored water chemistry solutions resulted in a longer lasting foam, by almost 1.5-3.0 times depending on the surfactant type. The bubble sizes were found to be much smaller in the tailored water chemistry when compared to high salinity water. Both longer foam-life and smaller bubble sizes are indicative of better, stronger and more stable foam. The foam rheology results also showed that the produced foams with the tailored low salinity water are of higher apparent viscosity when compared to those obtained with high salinity water.

This experimental study, for the first time, demonstrated the substantial improvements in the foam stability by using a tailored water chemistry. Such huge foam stabilization improvements obtained with tailored water chemistry has the promising potential to increase the apparent viscosity of injected gas and eventually EOR.

Integrated Stacked Carbon Storage in the Midcontinent of the United States

Authors:
Duguid, A., Battelle Memorial Institute
Hawkins, J., Battelle
Fukai, I., Battelle Memorial Institute
Middleton, R., Los Alamos National Laboratory
Hoover, B., Los Alamos National Laboratory
Swanson, J., Nebraska Public Power District

The Integrated Midcontinent Stacked Carbon Storage Hub (IMSCS-HUB) is studying the feasibility of integrated commercial CO_2 storage in Nebraska and Kansas. The objective of the project is to develop a stand-alone commercial storage project by 2025. The confluence of the regionâ€™s agricultural (corn) output, ethanol production, and oil industry make it an ideal candidate for early implementation of a stacked-CO_2-storage hub.

Carbon capture, utilization, and storage in the Midcontinent will set the example for future integrated projects by combining CO_2 from multiple, varied, sources with storage in stacked saline formations and CO_2-enhanced oil recovery reservoirs allowing shared transport and surface infrastructure. The low cost and high readiness level of capture from ethanol plants, coupled with regional storage capacity means that the midcontinent is poised to take full advantage of the CO_2tax credit in section 45Q of the US tax code. However, careful planning for pipeline routing and infrastructure development will be required to allow existing electric utilities to bring CO_2 to market as post-combustion capture becomes commercial. The addition of captured CO_2 from electric utilities would significantly increase capture and storage in the study area, from approximately 3 million tonnes per year to approximately 10 million tonnes per year. Coal-fired utilities in the Midcontinent area produce some of the lowest cost power in the United States and are likely to produce power for decades to come.

This presentation covers the process to select possible CO_2 sources for inclusion in the IMSCS-HUB including development of transportation infrastructure and routes considering both early ethanol and later electric utility source locations, along with estimated capture and transportation costs. The presentation also covers the process to identify and characterize stacked-storage sites in the region, including the estimation of storage capacities, storage costs, and methods of integration with CO_2 sources and transportation infrastructure.

Mechanical Earth Modeling at Farnsworth Unit, Texas, USA: Implications for Caprock Integrity, Fault Reactivation, and Reservoir Forecasting

Authors:
Balch, R., New Mexico Institute of Mining and Technology
Will, R., New Mexico Institute of Mining and Technology
McMillan, M., New Mexico Institute of Mining and Technology

The Southwest Partnership on Carbon Sequestration administers a CO_2 carbon capture utilization and storage (CCUS) project sponsored by the U.S. DOE. The project is located in a mature waterflood converting to enhanced oil recovery using anthropogenic CO_2. Major project goals are optimizing the storage/production balance, ensuring storage permanence, and developing best practices for CCUS. A key component of this effort is generating detailed geologic and flow models.

This paper presents the development of a 3D coupled Mechanical Earth Model (MEM) for use in assessment of caprock integrity, fault reactivation potential, and stress dependent permeability in reservoir forecasts. Mechanical property estimates computed from geophysical logs at selected wellbores were integrated with 3D seismic elastic inversion products to create a 3D â€œstaticâ€ mechanical property model sharing the same geological framework as existing reservoir simulation models including three major faults. Stresses in the MEM were initialized from wellbore stress estimates and reservoir simulation pore pressures. One way and two way coupled simulations were performed.

Coupled simulations were performed on history matched primary, secondary, and tertiary recovery periods, and on an optimized WAG prediction model. These simulations suggest that the field has been operating at conditions which are not conducive to caprock failure or fault reactivation. Two way coupled simulations were performed in which permeability was periodically updated as a function of volumetric strain using the Kozeny-Carmen porosity-permeability relationship. These simulations illustrate the importance of frequent permeability updating when recovery scenarios result in large pressure changes such as in field re-pressurization through waterflood after a long primary depletion recovery period. Conversely, forecasting is less sensitive to permeability update frequency when pressure cycles are short and shallow (WAG).

Funding for this project is provided by the U.S. Department of Energy under Award No. DE-FC26-05NT42591. Additional support has been provided by site operator Perdure Petroleum LLC.

Miscibility Management for CO2 EOR and Storage at Depleted Oil Reservoirs

Authors:
Chen, P., University of Houston
Thakur, G., University of Houston
Balasubramanian, S., The University of Houston

Most reservoirs have experienced pressure depletion below original bubble point pressure during primary production, which are then repressurized with waterflood. To maximize the utilization of anthropogenic CO_2 for miscible EOR in these depleted oil reservoirs, we investigated the effect of impurity of anthropogenic CO_2 stream and production history of depleted reservoir on the reservoir fluidsâ€™ characteristic, especially with regard to miscibility with the injectants.

The reservoir fluid phase behavior studies were carried out to evaluate the potential of applying miscible CO2 flooding in depleted reservoirs, taking NHK reservoir (in Assam of India) as an example. The rising bubble apparatus (RBA) and slim tube test were employed to determine CO2 MMP. The CO2 MMP with the reservoir fluid were monitored as it produced gas through a differential liberation procedure. The CO2 stream with impurities of methane and ethane was investigated to determine their sensitivity on the CO2 miscibility. PVT simulation study were then carried out to validate the laboratory tests.

In NHK reservoir, the MMP of CO2 with live oil before pressure depletion was determined to be above 4000 psi. After loss of dissolved gas through multiple-step differential liberation procedure, the MMP was dramatically reduced. The MMP measurements on the deferentially liberated oils demonstrate that this process actually lowers the MMP and hence is beneficial to the miscible gas flood. Moreover, CO2 with 5% lean gas contaminants (4.5% C1+0.5% C2) only slightly altered the MMP. When C1+C2 content increased to 10%, the CO2 MMP increased significantly. This indicates a negative effect exists when either CO2 is mixed with trapped lean gas or the anthropogenic gas purity is low. In all, this study provides a strategy for optimizing the MMP of a CO2-crude oil mixture in a depleted reservoir so that CO2 can be utilized more effectively in a miscible displacement process.

Calcination/Carbonation Cycles on SO2 & CO2 Co-Capture By Two Structurally Distinct Limestones

Authors:
Homsy, S., KAUST
Dibble, R., KAUST

Heavy fuel oil (HFO) is a low cost, highly sulfurous, unavoidable residual petroleum distillation byproduct representing 2-5% of refinery yield. Although phasing out HFO usage as a power source is unlikely while the world remains reliant on petroleum products, literature on CCS implementation at HFO-fired power plants is scarce. Calcium looping (CaL) is an advanced post-combustion CO_2 capture technology that utilizes limestone-derived calcium oxide as a CO_2sorbent. Since CaO sulfation occurs at typical process conditions, CaL can simultaneously concentrate CO_2 and capture SO_2 from HFO flue gas. CaL typically utilizes two interconnected fluidized beds to cycle between: (1) CO_2capture by CaO carbonation and (2) CaO regeneration by $CaCO_3$ calcination. As limestone is cycled, its capture capacity decays. The rate of decay is dependent on sorbent exposure to steam and deactivating impurities such as SO_2, which are present in HFO flue gas. In this study, samples of two structurally distinct limestones were cycled in the presence of: (1) steam, (2) both steam and SO_2. Relative limestone attrition, sintering, and deactivation rates were assessed. Samples at different stages of activity (X_{max} = 0.6, 0.3 and 0.1) were collected and used to explore the influence of sorbent cycling and $CaSO_4$ content on CO_2 and SO_2 co-capture from synthetic HFO flue gas during carbonation. Bubbling fluidized bed (BFB) reactors equipped with effluent line sampling for continuous CO_2, SO_2, and O_2 gas analysis were used. Sulfation and carbonation of cycled material was further explored by means of TGA analysis. Porosimetry and BET surface area analyses were performed on samples of the cycled limestones, and differences in CO_2 and SO_2 capture performance were correlated to differences in material surface structure and porosity.

Development of CO2 Capture Technologies for Coal-Fired Power Plants in China: Idealized Test Challenges for China-Specific Conditions

Authors:
Hao, P., China Energy
Liu, X., China Energy
Wang, B., National Institute of Clean-and-Low-Carbon Energy
Cui, Q., National Institute of Clean-and-Low-Carbon Energy
Zhao, X., National Institute of Clean-and-Low-Carbon Energy
Xu, B., China Energy
Singh, S., NICE America Research, Inc
Ku, A., China Energy
There are important differences in the operation of power plants in China that affect the performance and economics of post-combustion CO_2 capture systems relative to other parts of the world. These include the possibility of lower capture rate requirements, salient differences in the installed power plant infrastructure, and flexible operation in the forms of reduced capacity factors and load-following. This paper quantifies these China-specific conditions that CO_2 capture technologies will encounter at coal-fired power plants in China and proposes a suite of idealized test challenges that can be used to supplement testing of candidate technologies during the pilot stage. Data from these challenges can be used to improve accuracy in estimates of operational performance, robustness, and economic potential and also provide early insight into technology development needs.

New Amine-Based Membranes for Post- and Pre-Combustion CO2 Capture

Authors:
Ho, W. S. W., The Ohio State University

This presentation covers new advances in amine-containing membranes for post-combustion CO_2 capture from flue gas in coal- and/or natural gas-fired power plants and pre-combustion CO_2 capture from synthesis gas derived from coal and/or natural gas. We have synthesized highly CO_2-selective membranes comprising fixed-site and mobile carriers, involving the facilitated transport mechanism based on reversible CO_2 reactions with amine carriers. The membranes remove H_2S even faster than CO_2. In general, the membranes need to be tailor-made and tuned specifically for those applications. For example, post-combustion carbon capture requires a high CO_2/N_2 selectivity of 140 together with a very high CO_2 permeance of greater than 700 GPU in order to use a stand-alone membrane process. We have synthesized composite membranes showing a high CO_2 permeance of about 1500 GPU and a high CO_2/N_2 selectivity of more than 180, which are well above the 2008 Robeson upper bound. The membrane was scaled up successfully to 14 inches in width, and it was fabricated into spiral-wound membrane modules up to a membrane area of about 3 m^2. Three modules each with 1.4 m^2 were tested at NCCC, and they showed the same performance as those tested at OSU using simulated flue gas, including excellent membrane stability. The modules are promising for meeting the DOE cost target set for 2025. On the other hand, pre-combustion carbon capture demands the membrane with a very high CO_2/H_2 selectivity of 100 along with a modest CO_2 permeance of about 200 GPU. We have synthesized composite membranes exhibiting 206 GPU CO_2 permeance with 103 CO_2/H_2 selectivity. Based on the synthesized membranes with tuned H_2S/CO_2 selectivities, a one-stage membrane process for pre-combustion carbon capture has been proposed and designed, indicating 6 ppm H_2S in the H_2 product achievable and an attractive increase in the cost of electricity (15.66%).

Transformational Sorbent-Based Process for a Substantial Reduction in the Cost of CO2 Capture

Authors:
Jain, R., InnoSepra, LLC

InnoSepra has developed several physical sorbent-based CO_2 capture technologies that can substantially reduce the cost of CO_2 capture from a variety of sources. The physical sorbents used in the InnoSepra process need about 58% lower absolute amount of energy (GJ/MT of CO_2) compared to MEA and since this energy is needed at about 100°C compared to >150°C for MEA and other solvents the impact of CO_2 capture on power loss is at least 70% lower. The first generation InnoSepra CO_2 capture technology, funded by a DOE project, was demonstrated at pilot scale at a coal-fired power plant. It demonstrated more than 94% CO_2 recovery, more than 99% CO_2 purity, more than 40% reduction in the capital cost, and more than 60% reduction in the steam extraction from the power plant leading to more than 40% reduction in the CO_2 capture cost. The second generation InnoSepra process tested at semi-bench scale can provide more than 50% reduction in capital cost (excluding CO_2 compressor), and more than 75% reduction in thermal power needed leading to more than 50% reduction in the capture cost. This process will be further tested at the pilot scale under a DOE-funded project and has the potential to capture CO_2 at $30-35/MT with little or no increase in LCOE (Levelized Cost of Electricity). In additional to these technologies InnoSepra will present its work on flue gas purification that is applicable to all CO_2 capture technologies. For solvent-based technologies it can lead to more than 80% reduction in solvent loss due to degradation and due to aerosol emissions. For non-solvent technologies it can lower the cost of CO_2capture equipment. InnoSepra will also present its work on CO_2 capture from low concentration sources (1-4% CO_2)which is directly applicable to CO_2 capture from NGCC plants.

Application of Hydrodynamic Flow Modeling Integrated with Geomechanics for Partially Depleted Sandstone Reservoir: Fwu CO2-EOR Project

Authors:
McMillan, M., New Mexico Institute of Mining and Technology
Will, R., New Mexico Institute of Mining and Technology
Ampomah, W., New Mexico Institute of Mining and Technology
Balch, R., New Mexico Institute of Mining and Technology
Czoski, P., New Mexico Tech

This work utilizes a full-field, history-matched, compositional hydrodynamic coupled geomechanical model for assessing the Farnsworth Field Unitâ€™s (FWU) active CO_2-WAG project. Depletion impacts the volumetric strain which intern affects the permeability of the formation. These geomechanical impacts are accounted for by using the Kozenyâ€“Carman relationship, applied at varying coupled frequencies and compared to uncoupled simulations. These permeability updates can affect the volume of CO_2 sequestered and the safety and security of the stored CO_2.

A static mechanical earth model (MEM) which incorporated the existing Southwest Regional Partnership (SWP) geological model, available geophysical logs, 3-dimensional seismic elastic inversion properties and lab derived strength and failure criteria was constructed for use in transient coupled geomechanical model. Appropriate overburden, under-burden, and side-burden formations required to impose mechanical boundary conditions are also included. The Kozenyâ€“Carman relationship coupled at various frequencies, are utilized to perform hydrodynamic geomechanical simulations of primary, secondary (water-flooding) and tertiary (CO_2 WAG) recovery periods. These results are used to investigate the range of geomechanical outcomes. Volumetric strains and effective stress perturbations are assessed along with diagnostic tools such as p-q stress paths and Mohr circles with associated failure envelopes. Fault parameters such as slip tendency and distance to failure are also presented. These form the basis for a discussion of appropriate geomechanical modeling criterion for the Morrow B.

The Kozenyâ€“Carman relationship has been applied to the Morrow B coupled compositional hydrodynamic geomechanical simulation. Traditional geomechanical diagnostic procedures: Mohr Coulomb circles and p-q stress paths with associated failure lines, fault closeness to failure and slip tendency have been used to illustrate our findings.

This is the first study of coupled hydrodynamic-geomechanical simulation of Morrow B CO_2-WAG project and provide an assessment of caprock integrity and mechanical fault stability.

CMTC-554157-MS

CO_2 Sequestration - Enhanced Oil Recovery - CO_2/EOR "Huff-n-Puff" in the 21st Century

David Merchant, Merchant Consulting

This paper was prepared for presentation at the Carbon Management Technology Conference held in Houston, Texas, USA, 15–18 July 2019.

Abstract

During the past 47 years, CO_2 flood technology for Enhanced Oil Recovery projects evolved from a partially understood process filled with uncertainties to a process based on proven technology and experience. Many questions involved with CO_2 flooding have been thoroughly analyzed and answered. This knowledge is currently being used by a limited number of companies that actually know how to **design**, **implement**, and **manage** a CO_2 flood for long term **profit**. The purpose of this report is to help disseminate this knowledge to operating companies interested in EOR flooding or to CO_2 Sequestration Communities interested in storing CO_2 in EOR projects.

Carbon Sequestration Options for CO_2/EOR from the 20th Century:
In 2015 and 2017, Merchant Consulting published two papers: **CMTC-440075-MS "Life beyond 80 – A look at Conventional WAG Recovery beyond 80% HCPV Injection in CO_2 Tertiary Floods"** and **CMTC-502866-MS "Enhanced Oil Recovery – The History of CO_2 Conventional WAG Injection Techniques developed from Lab in the 1950's to 2017"**. The objective of these reports were to target all **"10 CO_2 Recovery Methods"** used to date including **"Conventional WAG Techniques"** Results show Tertiary Oil Recovery can be as high as 26% OOIP with 190% HCPV Injected.

Carbon Sequestration Options for CO_2/EOR in the 21st Century:
From 2005 to 2019, the **Shale Oil Revolution** killed CO_2/EOR Sequestration by driving oil prices down to the \$50-\$60 dollar range along with the inability to lower CO_2 Capture costs. The purpose of this report is to provide the CO_2 Sequestration Community an update of **combining two existing Conventional CO_2/EOR Recovery Techniques** that will compete economically with today's **Shale Oil Revolution (Primary Recovery - Single Cycle)**. This includes:

"Shale Oil CO_2/EOR Huff-n-Puff" Oil Recovery to recover 2X to 3X times todays 1X Recovery.

New methods also target: Zero Emissions (No Flare) and incorporates Q45 Tax Credits

Introduction – CO_2 Flooding

CO_2 Flood History

The Permian Basin has had over 95+ years of oil production history (1923-2019) and produced over 35 BBO (Billion Barrels of Oil), 80 TCF (Trillion Cubic Feet) of gas, and represents the largest petroleum producing area in the United States. The Permian Basin covers 75,000 square miles in West Texas and Southeastern New Mexico. Its first gusher was drilled in 1923. In 1972, CO_2 injection was first introduced commercially in the Sacroc Field. Today, over 1.3 billion barrels have been produced with CO_2 and accounts for over 350,000 BOPD from over 17,000 production wells and 14,000 CO_2 injection wells, through 4,500 miles of pipeline.

Carbon Sequestration Options for CO_2/EOR from the 20th Century (Lessons Learned)

In 2015 and 2017, Merchant Consulting published two papers: **CMTC-440075-MS "Life beyond 80 – A look at Conventional WAG Recovery beyond 80% HCPV Injection in CO_2 Tertiary Floods"** and **CMTC-502866-MS "Enhanced Oil Recovery – The History of CO_2 Conventional WAG Injection Techniques developed from Lab in the 1950's to 2017"**. The objective of these reports were to target all **"10 CO_2 Recovery Methods"** used to date including **"Conventional WAG Techniques"** which have been used in over 90% of all the EOR projects conducted to date. Results show Tertiary Oil Recovery can be as high as 26% OOIP with 190% HCPV Injection of CO_2.

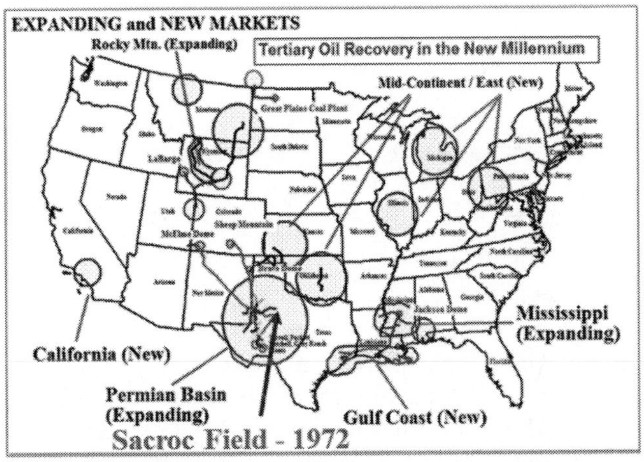

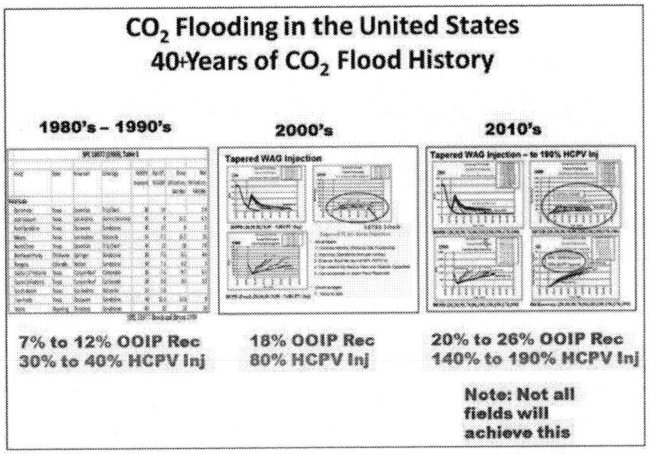

Unconventional Shale Oil Recovery in the 21st Century (United States Today)

The introduction and economic impact of **Unconventional Shale Oil Recovery** into the Permian Basin, Eagleford, and Bakken regions of the United States has been a game changer for many regions around the World. According to the American Petroleum Institute, the U.S. Crude Oil Production reached an all time high of 12.2 MMBOPD (million barrels per day) in May, 2019, with Texas Crude Oil output exceeding 5 MMBOPD for the first time. In addition, U.S. exports reached 8.1 MMBOPD, an increase of 10.5% over May, 2018.

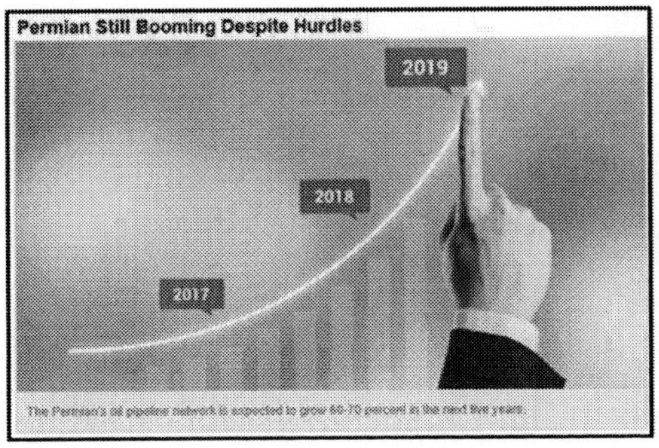

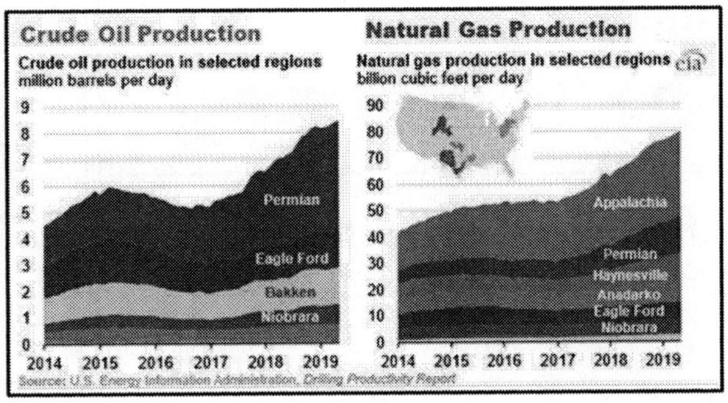

Shale Oil Revolution (The Good ....)

From a **United States** perspective, the U.S. currently holds the title of **Global leader in Recoverable Oil Resources**, according to the latest annual report of world recoverable oil by energy research firm Rystad Energy, a 20 percent increase from previous stated official reserves. With 293 billion barrels of recoverable oil resources, the U.S. beats out both Saudi Arabia and Russia by 20 billion barrels and 100 billion barrels, respectively. According to Rystad analysis, the Permian's tight oil plays hold 100 billion barrels of recoverable oil resources and the resources there remain largely flat from the previous year. It was also noted that production was not fully replaced by increased reserves in some U.S. shale plays, including Eagle Ford in Texas and Utica in Ohio. With the stabilization in oil prices, oil companies have been focusing more on core development and cash flows rather than new exploration.

In addition to the small players (EOG, Pioneer Resources, etc.), Major Oil Companies have also entered into the game. Chevron said it had plans to double its production to the equivalent of 900,000 BOPD by 2023, and Exxon sait it planned to produce 1 MMBOPD (one million barrels per day) by 2024, an 80% increase over its current level.

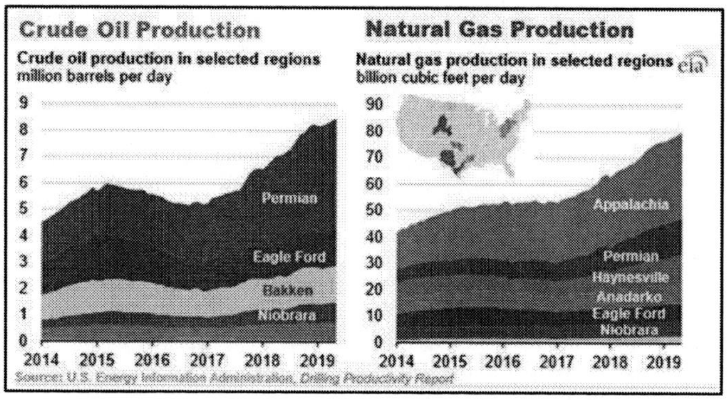

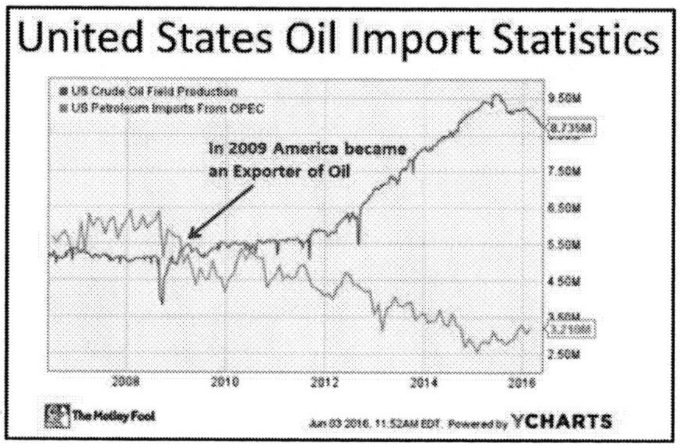

Drilled but Un-Completed Wells (DUC's)

In addition to reaching record breaking levels, the Permian Basin along with the remaining six regions contains a record number of DUC's (Drilled but Uncompleted Completions). These wells provide a "Nest" full of locations already drilled, but waiting to be completed, providing that safety net to fall back into when oil prices fall. Since this technology provides a quick entrance back into the Oil Market, the response time utilizing DUC's is much better and quicker to fill the Energy Gap between Supply and Demand when compared to other recovery methods.

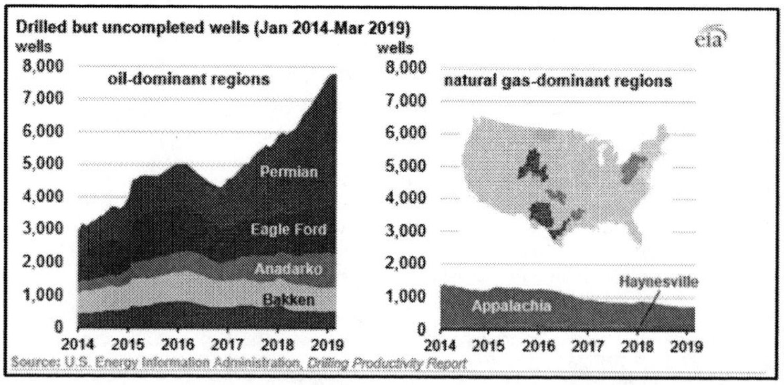

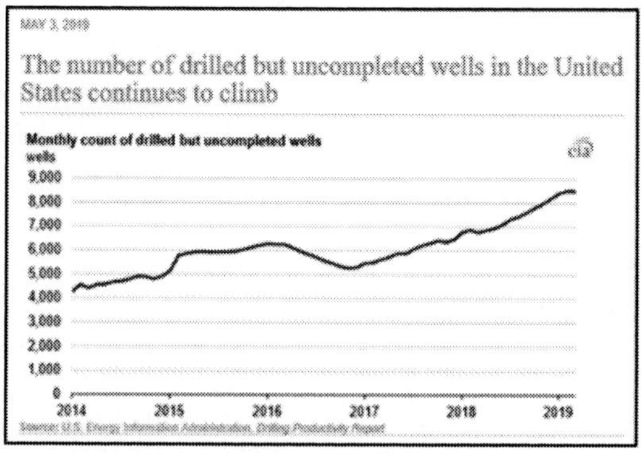

The number of drilled but uncompleted wells in seven key oil and natural gas production regions of the United States has increased over the past two years, reaching a high of 8,504 wells in February, 2019, according to the EIA's Drilling Productinvity Report (DPR). The most recent count, at 8,500 wells in March 2019, was 26% higher than the previous March in 2018.

Shale Oil Revolution (The Bad ....)

From a **Total World perspective**, this impact of the **Shale Oil Revolution** has been felt all over the World. This is shown below by Baker Hughes Well Rig Count reports over the period 2013 to 2016. In 2015, Total Rig Count dropped on a Global Basis, impacting the total Global Market. In 2019, WTI oil price settled in between $50 to $60 dollars per barrel. This impact also had an effect on North Sea Carbon Capture efforts to recover oil not recoverable under Primary Recovery. As of 2019, North Sea Primary Oil Recovery has recovered 80% of its ultimate Primary Oil Potential. Tertiary CO_2 Recovery would have saved the Oil Industry in the North Sea. Today, decommissioning will plug 4,000 wells, along with the removal of 300 platforms, and 20,000 Km of pipelines at a cost of 35 Billion Pounds.

Total Globe Perspective- Rig Count

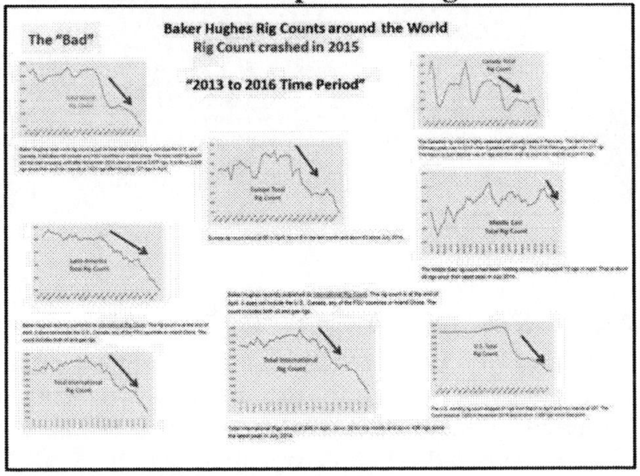

North Sea Abandonment

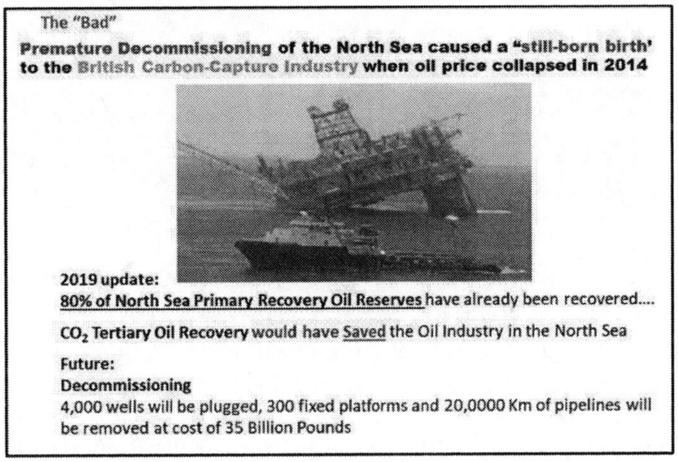

Shale Oil Revolution (The Ugly ....)

Flare History across America

In every new basin of the World, the amount of associated gas not available for market is flared. This has been true across the globe including: Middle East, North Africa, North Sea, Alaska, Gulf Coast, Africa, and South America. As shown below, the prolific waste from the Permian Basin and Bakken is no exception. This occurs when rising oil production outstrips processing plant and pipeline production capacity, thus the associated hydrocarbon gas produced with the oil is flared.

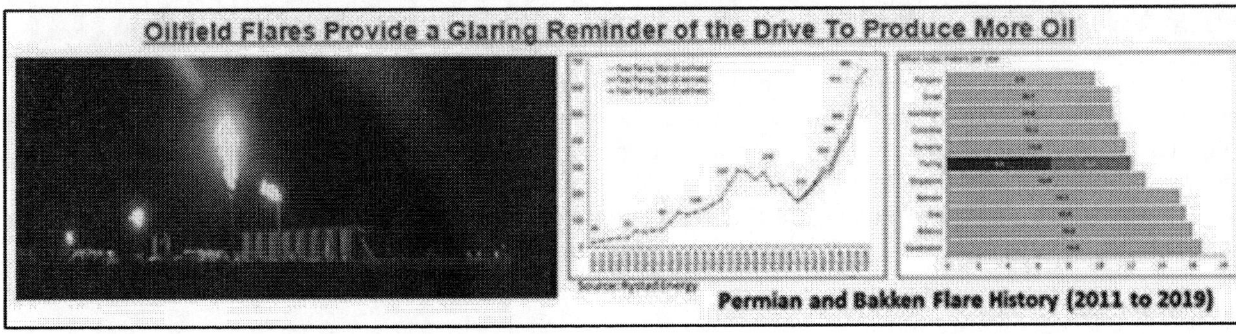

The volume of natural gas flared in the Permian basin MMCF/day (million ft3/day) has raised sharply since early 2017. When you compare the total amount of Gas Flared from the Permian and Bakken, Basins, more than half of the total comes from the Permian, which has surpassed the Bakken by more than tripling the volume of gas burned per day since early 2017. The severity of the problem, though, is more acute in the Bakken. While Rystad said about 5% of the Permian gas is flared, in the Bakken, 19% of the gas is flared. The Permian produces far more gas but has the advantage of being close to growing markets for it—chemical plants, export pipelines to Mexico, and natural gas liquefaction facilities—while North Dakota is a distant option for Midwestern markets well supplied by other plays.

21st Century Un-Conventional CO_2 Tertiary Recovery Methods

CO_2 Storage Solutions
Permian Basin Goal – Zero Gas Emissions

CO_2 Plant Capture | **CO_2 Pipeline Infrastructure** | **CO_2 Reservoir Storage and Sequestration**

CO_2 Storage Solutions – CO_2 Plant Capture

Electrical Power across the Permian Basin (A Solution to the Flare Gas Problem)

In addition to Flaring natural gas, ensuring adequate power for Permian oil field operations has been increasingly challenging. Operators are seeing longer lead times to get power, higher cost of infrastucture and power outage issues. The Electric Reliability Coucil of Texas (EPCOT) reported in July, 2018 that West Texas power use increased by nearly 250 meagawatts over the previous year due to increased oil and gas activity and is trending upward. On May 10, 2018, power use in West Texas exceeded 3,400 megawatts (3.4gigawatts) for the first time. (One megawatt powers 200 homes, and one gigawatt = one nuclear power plant or 680,000 homes.

The **Solution** to this problem is to build more Power Plant Capacity. The question becomes **Which Type**?

Century Plant Solution

Gulf Coast Pipeline Solution

Shovel Ready Coal Projects

Petra Nova Coal Solution

NetPower Gas Solution

Hydrogen Solution

1The Century Plant currently provides CO_2 to the Permiabn Basin capable of receiving Q45 credits.
2. A massive pipeline program in Gulf Coast to export LNG overseas to replace Coal-fired Power Plants.
3. The Permian pursued two Coal Powered Power plants (Odessa and East Permian) but both failed economically.
4. The Petra Nova Plant (Southwest of Houston) is Coal Powered and was built "On Time and On Budget".
5. The NetPower **Hc-gas Powered Power Plant** utilizes both "Produced and Flare Gas" to make CO_2 and Water.
6. In the future, the Energy Gap between Consumption and Supply will be moving towards the "Hydrogen Era"

Comparison Results:

Currently the Oil Industry is moving towards eliminating the flare by moving Hc-Gas, LNG, and Oil out of the Eagleford and Permian Basins through a massive pipeline investment to the Gulf Coast and Over Seas. The overall impact is good when considering we are helping mitigate Climate Change by closing down Coal Powered Coal Plants abroad. When comparing **Coal Capture projects** identified above for the Permian Basin, they all create CO_2; but, do not do anything to eleiminating the flare problem. The NetPower option utilizes Hc-gas and Oxygen to create both CO_2 and water, plus electricity and provides the best option in today's market to solve the flare problem and make CO_2/EOR Unconventional oil recovery in the Permian with "Zero CO_2 Emissions."

Permian Basin Unconventional (Oxy Permian Lab and Well Simulation Results)

Oxy Permian - Unconventional CO_2 Learnings

What is Huff-n-Puff?

The Huff-n-Puff Recovery process is a single-well process that has been successfully implemented in conventional reservoirs for several decades. Since the recovery process is a miscible process, the injecting fluid can be Methane (CH_4), Ethane (C2), Propane (C3), CO_2, H_2S, or a combination (C2+C3) of components. For example: Oxy Permian included in its modelling efforts a comparison between pure CO_2, Methane (CH_4) only, and Separator Gas (C2and C3+) injection in its model predictions.

CO_2 Huff-n-Puff for an individual well is a three-step process
Step 1: CO_2 is injected in the production well.
Step 2: The well is shut in for a predetermined time called the "soak period."
Step 3: The well is returned to production, and oil flows toward the wellbore because of a pressure sink.

The reason why CO_2 Huff-n-Puff works is simple "The presence of hydraulic and natural fractures provides a large **contact surface area** for injected gas to penetrate into the ultra-low permeability matrix. The main recovery mechanisms are: 1) vaporization of lighter oil components, and 2) interfacial tension (IFT) reduction at pressures above the minimum miscibility pressure (MMP). Below the minimum miscibility pressure (MMP), CO_2 improves oil recovery through the mechanisms of oil phase swelling, viscosity reduction, and gas-oil displacement. At pressures above the MMP, miscibility results in high displacement efficiency due to ultra-low capillary forces.

Oxy Permian in a recent publication stated "our study indicates carbon dioxide (CO_2) Huff-and-Puff **may** be a technically feasible EOR method for unconventional reservoirs." They make their claims based on the following:

Miscible Enhanced Oil Recovery (EOR) mechanisms in unconventional, tight reservoirs may be substantially different than those associated with miscible gas flooding in conventional reservoirs. Due to potentially low cross-well throughput in ultra-low permeability rock, the **Single-well Huff-n-Puff** process is being considered within the industry for application in unconventional reservoirs.

Reservoir Simulation (Simple Mechanistic Reservoir Modelling)

Compositional Reservoir Modelling provides the ability make future predictions under various operating scenarios beyond Lab and Core data. For example: Amoco Production in 1986, modelled the Conventional WAG process with a single 5-spot model under various scenarios to understand the CO_2/EOR process better. The result was Tapered WAG Injection.

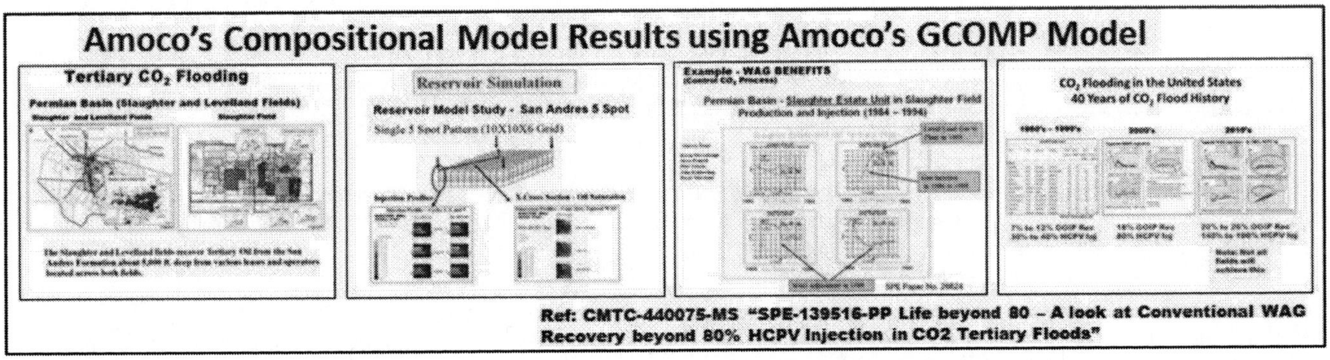

In a similar fashion, Oxy Permian collected fluid and rock samples from the Permian Basin Wolfcamp formation and tuned their EOS models. The results of both the lab fluid and rock data as stated by Oxy Permian demonstrated the Huff-n-Puff process could recover significant additional oil with additional cycles for Permian organic rich shales. The multi-cycle incremental recovery – even at the small core plug scale- suggests the significant potential for multiple huff-n-puff cycles for a future Unconventional EOR project design. In addition, for the Wolfcamp formation, CO_2 was found to be the best solvent for Tertiary Oil Recovery when compared with other injection gases. In addition, since CO_2 is a Multiple Contact Miscible process with the oil, the fluid composition of the produced oil changes with time.

Reservoir Simulation (Oxy Permian's Lab Analysis: Fluid and Core data)

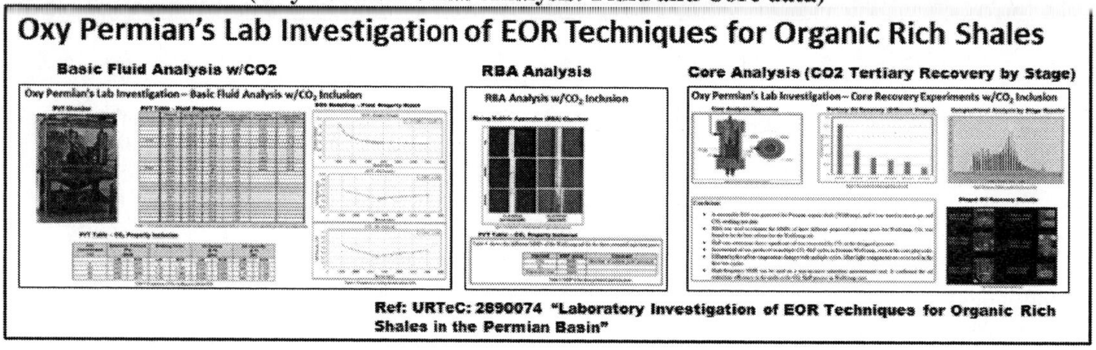

Reservoir Simulation (Oxy Permian's Single Well Model Predictions)

Compositional Reservoir Modelling is predicated on the ability to properly match field historical performance and lab data. The compositional model developed by Oxy Permian utilized an 11 component description based on the Peng-Robinson Equation of State (PR EOS). The primary recovery history match (Base Case) was calibrated to match Wolfcamp primary performance using assisted history matching techniques. The cycle scheme for the non-optimized (base case) Huff-n-Puff forecast assumed a schedule of 10 days of injection at 1.0 MMSCFD, 7 days of soaking, and 180 days of production. **The incremental oil realized under the un-optimized base case was around 23% over a period of 2 years with a gross utilization of around 22 MSCF/Bbl.**

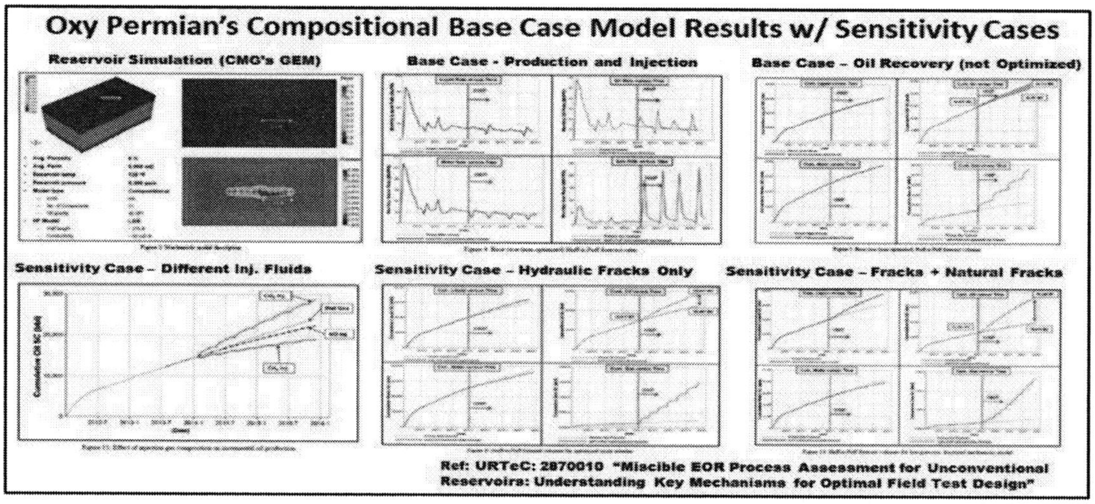

Cycle Scheme Optimization (Sensitivity Case Runs)

Optimizing the injection-soak-production design greatly influences CO_2 Huff-n-Puff performance. **The Optimized Huff-n-Puff oil recovery wedge was 81% higher than primary production (compared with 23% for the non-optimized case.** In addition, gross utilization calculated was 25 MMSCF/BBL).

Effect of Natural Fractures on Process Performance

The presence of hydraulic and natural fractures provides a large contact area for injected gas to penetrate into the low-permeability matrix to improve recovery. **The Huff-n-Puff wedge was 250% higher than primary production for the naturally fractured reservoir, compared to 81% higher for the optimized Huff-n-Puff hydraulically fractured case.** In addition, gross utilization was 18 MMSCF/Bbl for the naturally fractured case, compared to 25 MMSCF/Bbl for the base.

Effect of Injection Gas Composition on Process Performance

The effect of injection gas composition on incremental oil production was also evaluated. The injectants evaluated included CO_2, wet gas (CH_4 ~ 65 mole %, C2 ~ 15 mole %, C3+ ~ 20 mole %), and methane (CH_4). CO_2 had the highest recovery.

Conclusion

Oxy Permian <u>demonstrated</u> through lab and Reservoir Simulation, the Huff-n-Puff CO_2/EOR process could recover significant additional oil beyond Primary Unconventional Recovery in Permian organic rich shales.

Bakken Unconventional Shale Oil

As one of the largest unconventional resources in the World (200,000-square-mile-area in North America) with over 10,000 wells drilled in the past 10 years), the **Bakken Petroleum Basin** contains approximately 3.6 billion barrels of recoverable oil as estimated by the U.S. Geological Survey in 2008. The economic impact was seen when oil production grew significantly from 2009 to 2014 from 0.2 MMBOPD to 1.1 MMBOPD. In addition to the oil, a large volume of associated gas (1.6 BCF per day) was also produced. A substantial part (>10%) was flared off because of the low natural gas price and limited infrastructure for gathering and transporting the gas from well sites to processing plants.

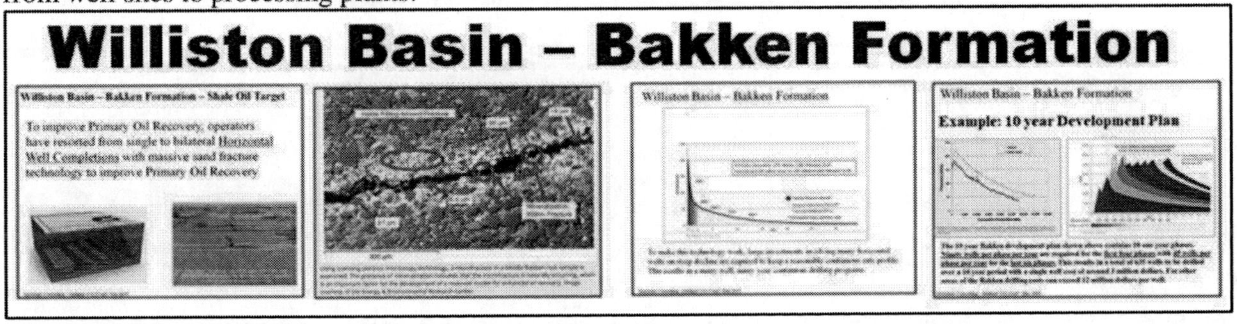

The Energy & Environmental Research Center (EERC) is currently focused on solving the Bakken formation's energy and environmental challenges. Through its Center for Climate Change & Carbon Capture and Storage, the EERC is engaged in research activities in direct support of carbon management. One significant carbon management effort led by the EERC is the Plains CO_2 Reduction (PCOR) Partnership. The PCOR Partnership is one of seven regional partnerships established by the U.S. Department of Energy National Energy Technology Laboratory to assess and develop carbon storage opportunities.

The main target of this drilling has been two of the non-shale low-permeability units, the Middle Bakken Member and the Three Forks Formation. Over the last 7-8 years, there have been a number of pilot tests for both water and gas injection. Results showed CO_2 and produced Bakken gas to be miscible with the oil at reservoir conditions (>5000 psi, 230°F), and measured MMPs for pure CO_2 and ethane were 2528 psi and 1344 psi, respectively. Injectivity of gas or water does not appear to be an issue in the Bakken; however, the projects, in general, show early breakthrough times and poor reservoir sweep efficiencies. Although the production rate is encouraging, the recovery factor (averaging 7% per well) is small compared to the huge amount of oil in place. How to improve oil recovery becomes a critical step toward future development of the basin.

Lessons Learned -Confinement and "Frac Hits"

Vertically, confinement in the Bakken, and elsewhere, has so far not been a big concern to operators, but **Horizontally** with **"Frac Hits"**, we have a bigger problem. While a large surface area for the gas to interact with is critical to achieving success, there are cases where the tensile fracture networks created during fracturing, along with natural fracture networks, could be overly extensive, allowing the gas to spread quickly into adjacent wellbores, thus lowering ultimate oil recovery.

Horizontal Confinement is not a new issue. These issues have been a part of Conventional CO_2 flooding since its conception. One of the best examples involves the Central Mallet Unit in Slaughter field. (Ref: SPE 26624 "Reservoir Management in Tertiary CO_2 Floods', published in1993). In fields with multiple operators and leases, **"Water Curtain Technology"** was first introduced to **contain** the CO_2/EOR flood within unit boundaries by incorporating water injectors at lease boundaries. But in the Central Mallet Unit and Slaughter Estate Units, Chichenleg pattern producers were converted to water injection, thus increasing reservoir pressure above MMP across the reservoir, allowing the WAG injectors freedom to do what they were designed for; that is, managing conformance across the injection interval with water, rather than the injectors doing both conformance control and maintaining pressure above Minimum Miscibility Pressure (MMP). For the Bakken, water could work, but other alternative approaches and (Containment Area Solutons) need to be investigated.

Eagleford Unconventional (Proven Hydrocarbon Gas Huff-n-Puff Projects)

CO_2 Injection is not the only injectant to recover additional Tertiary Oil. In the spring of 2016, Houston-based EOG Resources let it be known that its shale EOR program was boosting production from vintage horizontal wells in its Eagle Ford Shale asset in south Texas. It is commonly accepted today all of these projects rely on the **Huff-n-Puff** Injection process using **Natural Gas** as the injectant mobilizing tertiary oil. In a recent quarterly earnings statement, EOG said it continues to see "strong results" from around 150 EOR wells, more than a third of which were converted in 2018. Analysts and engineering consultants have found about 100 other wells in the Eagle Ford that several other operators have converted into huff-and-puff injectors as well.

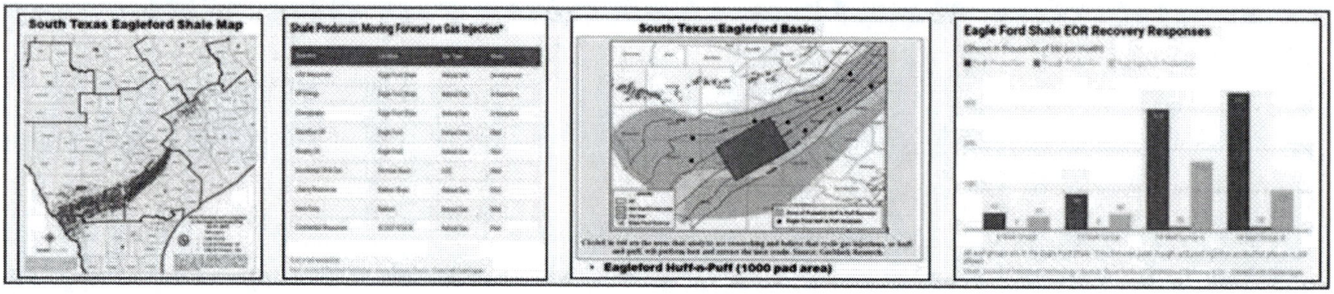

John Watson, a senior research analyst who put together a report late last year that highlighted production details of shale EOR projects, found dozens of pad wells that saw a combined 10-fold rise in production above their trough. Watson's report covers more than two dozen other shale EOR projects, though most lacked production results, revealing only project cost estimates. Among the standouts, a group of 11 wells that reached a combined peak production rate in December, 2011 of about 90,000 bbl a month. By August 2017, these wells were pumping out only 5,000 bbl a month. After gas injections began, the group produced 40,000 bbl a month—an average increase from about 15 BOPD to 117 BOPD per well. Another case involved 14 wells that peaked at 330,000 bbl a month in 2013, and then dropped to 10,000 bbl per month. Post injection, output increased to 170,000 bbl a month.

Pilot Response Proof:
A comparison of Oil Recovery for the different pattern areas is shown above, proving the Huff-n-Puff process works with actual <u>proven</u> results.

Confinement: Vertically, confinement in the Eagle Ford, and elsewhere, has so far not been a big concern to practitioners. "Horizontally, we have a bigger problem," said Chet Ozgen, a Reservoir Engineering Consultant with NITEC LLC, who explained that the worry here is over the long, tensile fractures that operators created via their early-generation stimulations. While a large surface area for the gas to interact with is critical to achieving success, there are cases where the tensile fracture networks could be overly extensive, allowing the gas to spread quickly into adjacent wellbores. In addition, Chet Ozgen has found that newer generations of hydraulic fracturing designs, ones aimed at generating near-wellbore complexity, are making for the best EOR candidates.

Sequencing and communication: The order in which gas injections take place between groups of wells is the key to optimizing the approach. And a big deciding factor here is how the pad wells communicate, which the shale sector has learned is almost always the case. Ozgen and others are using reservoir models to derisk the sequencing options. Based on his experience, Ozgen estimates that 50% of the scenarios where wells are highly communicative, which could be indicated by a history of intense frac hits during the completions phase, may turn out to be noneconomic. He compared this to converting isolated, individual wells to EOR, in which case the sequencing will have almost no impact on the economics.

In the worst cases, the big risk might be that the wells end up recycling the injected gas from one to the next without improving production. To sum up how critical sequencing is, Ozgen concluded: "If you do not find the correct order, you can lose money and your [incremental] recovery is horrible."

<u>Permian Basin and Eagleford:</u> What is the CO_2 Storage Capacity of a 1,000 Pad region (40 by 50 square mile area) located within either the Midland or Eagle Ford Basins Look Like?

Historically, **Scoping Study Model Analysis** has been used to make **Tertiary CO_2/EOR Predictions** in **Conventional CO_2 floods** for over three decades without having to conduct extensive Reservoir Modelling. Reservoir Modelling and analysis in the Unconventional Shale Oil Reservoirs has been primarily focused on understanding **Single Well Performance** or **Parent/Child Relationships** to better improve **Single Stage Primary Oil Recovery**. The question becomes?

What will the <u>Permian Basin</u> or <u>Eagleford</u> look like with the inclusion of CO_2/EOR Huff-n-Puff in the future by the year 2050?

The example below depicts a 1,000 PAD example (50 mile X 40 mile) grid or (area covering 2,000 square miles), that was used to compute CO_2 Storage and Capture values.

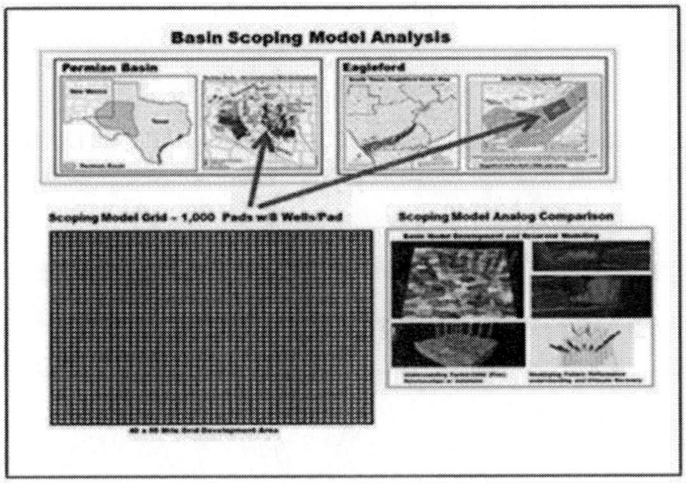

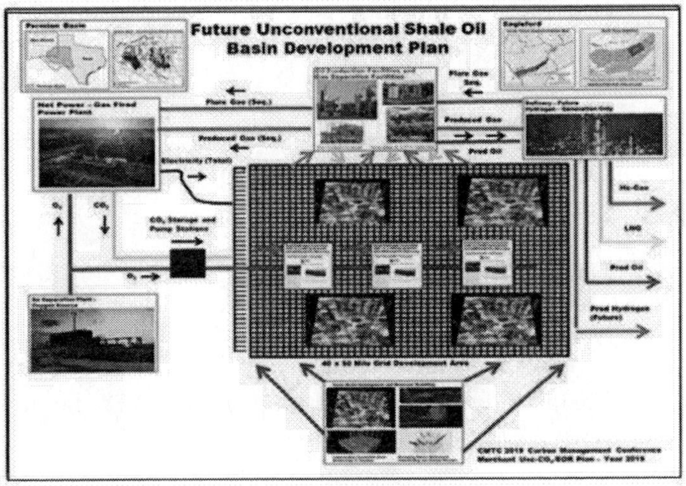

The Basin Area Scoping Model Prediction of CO_2 Storage and Sequestration was built on the following:

<u>Assumptions</u>

1. Assume that 1000 drilling PADS in the Eagle Ford or Permian Formation would be available for future CO_2 sequestration <u>or</u> 1,000 PADS would be located within the Wolfcamp zones A and B, which have similar CO_2 sequestering potential and CO_2/EOR Recovey characteristics, but over a much larger area in the Midland Basin.
2. Assume that each Drill Pad (PAD) was 1 mile by 2 miles in drainage area with 8 horizontal Eagle Ford wells per PAD
3. Assume that 5% of the active Huff-and-Puff (HnP) EOR gas injectant (in this case CO_2, not Rich gas) would be retained during 5 gas injection cycles per well
4. Assume that that 80% of the removed reservoir bbls could be refilled with sequestered CO_2 gas and the wells on each PAD would be abandon (with monitoring) to sequester about 4.5 Bcf per well PAD
5. Assume a conservative 1000 PADs (8000 wells) would be used for CO_2 sequestering at final abandonment (Fifth Cycle wth Huff, but no Puff (Storing CO_2 in Reservoir)

Results

The Scoping Model estimates 4.5 Tcf of CO_2 could be stored or sequestered in the ground permanently over a 50x40 square mile area (80 km x 64 km) of the Bakken or Midland Wolfcamp A and B reservoirs.

Conclusions

Technical Success Is Not Enough

No matter how inspiring or representative the early results appear to be, they have not proven to be enough to warrant major investments by most of the shale sector. Experts believe there are thousands of potential shale EOR locations in the Eagle Ford alone, yet only a relative small of projects have been piloted. Further, less than a dozen shale producers are known to be testing injection operations of various scales in south Texas. Some will rely on CO_2, such as Occidental Petroleum's Permian plans call for, but it appears the most popular approach will rely on natural gas.

The Question becomes "At what point will Unconventional Shale Oil CO_2/EOR make its mark in the 21st Century?"

Acknowledgements

I would like to express my gratitude to the members of the Carbon Management Technology Conference (CMTC) for allowing me the opportunity to be part of this CO_2 Conference. I would especially like to thank George Koperna of Advanced Resources International and Jose Figueroa for their support in making this effort possible.

In addition, I would also like to express my gratitude to Mike Uland and Jim Gilman of iReservoir in Denver, Colorado for opening my eyes up again to Oxy's recent efforts of expanding and implementing CO_2/EOR in the Permian Basin.

David H. Merchant
Merchant Consulting
Magnolia, Texas

Basin Study Analysis and Scoping Model Predictions
www.CO₂StorageSolutions.com

Reservoir Modelling and Management (VIP, Eclipse 300, GEM, MORE, Amoco's Gcomp Model, Montage-3d)
www.ReservoirManagementSolutions.com

CO_2 Seminars on Wheels (One or Two Day School) (When you can't make a Conference, then have the Conference come to **You**)
www.CO₂SeminarsOnWheels.com

Impact of NO2 co-Injection on Geological Carbon Sequestration

Authors:
Kumar, R., University of South Florida
Campbell, S. W., University of South Florida
Cunningham, J., University of South Florida
Geological carbon sequestration is a promising technology for large scale CO_2mitigation, especially from the point sources like power plants. The captured flue gas streams often contains small amount of impurities such as Sulphur oxides (SO_x) and Nitrogen Oxides (NO_x). In this paper, the objective is to study the impact of NO_2 co-injection on geological carbon sequestration in a deep saline carbonate formation. The objective of this research is accomplished by estimating the impact of NO_2 on following parameters a) changes in pH of the brine-rock-CO_2 system b) changes in mineralogy or porosity of the formation c) changes in solubility of CO_2in brine or solubility trapping.

TOUGHREACT 3.3 (developed by Lawrence Berkeley National Laboratory, Berkeley) has been used to simulate the brine-rock-CO_2-NO_2 system. The mineralogy of the carbonate formation is based on the Dollar Bay formation in South Florida basin. The simulated domain is comprised of 11 carbonate layers varying in terms of porosity, permeability, thickness and composition of carbonate minerals (mainly calcite and dolomite). Three different NO_2 mass fractions 0.15%, 1%, and 2% were co-injected with CO_2.

Results from the study suggests that co-injecting NO_2 acidifies the system to greater extent as compared to pure CO_2 injection. Higher degree of acidification causes more dissolution of carbonate minerals, leading to higher changes in porosity across the layers in case of NO_2 co-injection (magnitude in order of 10^{-3}), compared to pure CO_2 injection (magnitude in order of 10^{-4}). Solubility of CO_2remained largely unaffected with concentration of dissolved CO_2 for all four cases were evaluated around 50 g/kg of water. Based on the study, it can be concluded that co-injecting NO_2 can be a feasible option for geological carbon sequestration as it may render economic benefits by cutting the process cost associated with reducing NO_x concentration in the emission streams.

Pressure Interpretation for Leaky Well Characterization during CO2 Storage

Authors:
Mosaheb, M., Louisiana State University
Zeidouni, M., Louisiana State University
Depleted oil and gas reservoirs provide significant storage capacity for geological CO_2 storage. Leakage of the injected CO_2 from the target formation can adversely impact the groundwater resources and ultimately the surface environment. Well integrity is a key for secure underground storage. In this study, we provide a pressure interpretation method for leakage identification during CO_2 storage in a depleted reservoir in presence of multiple leaking wells. We investigate the multiphase flow of the injected and native fluids to interpret the pressure perturbation of the storage layer as well as the above zone. We provide an analytical pressure interpretation method to identify the existing wells that are acting as leakage pathways. The characterization method is capable to estimate the location of the leaking wells. This location identification is based on the traveling time of the pressure signals that are directly dependent on the traveling distance. Therefore, we can identify which wells are leaking among the numerous wells. In addition, the leakage rate and conductivity of the leaking wells are estimated. This step is based on the dependency of the pressure variations on the conductivity of the media. Results show that some leaking wells dominate the leakage pressure signals. This dominance is due to the less distance to the observation point and the greater leakage conductivity than the other leaking wells. We find an analytical approach to evaluate this leakage dominancy. The leakage dominancy approach is useful to estimate the optimum location of the observation points to enhance the value of the observed pressure variation and reduce the number of observation wells.

Options for Minimising CCS Retrofit Costs to Natural Gas and Coal Power Plants in US Markets

Authors:
Elliott, B., Bechtel
Gulen, J., Bechtel
Benz, G., Consultant
Gibbins, J., UK CCS Research Centre

The demand for â€˜plain vanillaâ€™ baseload CCS power plants in electricity markets seems to be waning. At the same time, very low, and possibly even negative, average tCO_2/MWh emission levels can be expected for future energy systems.

Since there is underutilised capacity in the national power plant fleet, a CCS deployment strategy based on retrofits is likely to be most cost-effective to meet these requirements, with two classes of retrofit of particular interest:

1. flexible natural gas power plants retrofitted with CCS; and
2. negative emission CCS plants, based on retrofits to coal plants converted to use biomass.

Flexible gas power plants retrofits could use post-combustion capture or hydrogen firing. The preferred option is not simply a matter of average load factors; there is a relatively wide overlap where project-specific factors will influence rational choice.

Biomass combustion in existing pulverized coal plants, cofired with coal or as a 100% conversion, is already a relatively mature technology. The nationâ€™s existing coal fleet therefore represents a huge potential both to produce negative emissions and to support employment in a new industry producing sustainable biomass.

For biomass plants, post-combustion capture is likely to be the preferred route, but again with a number of options available to confer benefits for owners through increased operational flexibility and future upgradeability.

Longer term deployment of CCS power plants retrofits will obviously need to be supported by a range of policy and market changes, not least a recognition of the unique services that CCS can offer. In the shorter term ways of building the first plants are required, using existing instruments such as 45Q as far as possible. Possible routes for reducing the costs of these initial projects, while still retaining their value as reference plants for future large scale rollout, will be discussed.

An Advanced Pccc Control Scheme Utilizing Predictive Control and System-Dependent Fast-Response Variables at the Uky-Caer 0.7 MWe Small Pilot Scale CO2 Capture Facility

Authors:
Pelgen, J. V., University of Kentucky
Irvin, B., University of Kentucky
Liu, K., University of Kentucky

With regard to post combustion CO_2 capture (PCCC) processes, the cost of capture is of vital importance to mitigating emissions. Economic implications present themselves if the capture system is often operated off of the best performance curve at given flue gas compositions and ambient conditions. Current process control methods involve the use of simple feedback controllers and proportionalâ€"integralâ€"derivative (PID) controllers. While these methods are industry proven for systems with low fluid inventory, improvements in control system response time for large fluid inventory systems, such as aqueous carbon capture units, can be made by using a form of feed-forward control that is based off of a process data generated model.

With the University of Kentucky Center for Applied Energy Research (UKy-CAER) PCCC process, there are many factors affecting system performance that are difficult to consistently control. These include: capture solvent concentration and CO_2 loading, system water balance, ambient conditions, solvent degradation and emission rates, and differences in manual responses. By utilizing a control system based model predictive control algorithm and selecting gas-side fast-response variables that are indicative of future system performance, these factors can be more precisely controlled. With conventional PID control using liquid-side variables, changes in system operation can only be realized after the solvent inventory has completely cycled. This leads to overshooting and undershooting energy requirements from 2-5% to reach 90% CO_2 capture. UKy-CAER has developed an approach for an advanced control scheme utilizing the CO_2 product flow rate at the stripper exhaust and real-time calculated solvent quality analysis to control the 0.7 MWe small pilot scale CO_2 capture process. Through use of a fast-response and predictive control approach, UKy-CAER can realize a 3-4x increase in the process state response time, and the specific reboiler duty requirement can be tuned to the real-time solvent quality.

CO2 Capture from Power Plant Flue Gas with the Mtr PolarCapTM Process: Update on Field Demonstration Tests

Authors:
Kniep, J., MTR
Baker, R., MTR
Casillas, C., MTR
Chan, K., Membrane Technology and Research, Inc.
Freeman, B., Electric Power Research Institute
Hofmann, T., MTR
Huang, I., Membrane Technology and Research
Merkel, T., Membrane Technology and Research, Inc.
Nguyen, V., Membrane Technology and Research, Inc.
Sun, Z., MTR
Wang, X., MTR

Membrane technology is an attractive option for CO_2 capture because of advantages such as energy-efficient passive operation, tolerance to acid gases and oxygen, a small footprint, no use of hazardous chemicals, and no steam use requiring modifications to the existing boiler and steam turbine.

Working with DOE, MTR has developed new membranes and process designs to recover CO_2 from power plant flue gas. MTR Polarisâ„¢ membranes have CO_2 permeances ten times higher than standard commercial membranes, which greatly reduces the cost of a membrane capture system. These membranes are combined with novel PolarCapâ„¢ process designs for high CO_2 removal rates (90%) or partial capture rates. Design calculations estimate that the PolarCap process can capture CO_2 at a cost as low as \$30/tonne under partial capture conditions, which is competitive with leading second generation capture approaches.

Between 2012 and 2015, MTR operated bench-scale (1 tonne CO_2/day) and small pilot (20 tonnes CO_2/day or 1 MW_e equivalent) systems at the U.S. National Carbon Capture Center (NCCC) in Wilsonville, AL. Overall, the systems accumulated over 11,000 hours of run time, demonstrated stable membrane performance and validated prototype module designs and advanced membranes under real world conditions.

After testing at NCCC, the small pilot membrane system was moved to Babcock & Wilcoxâ€™s (B&W) boiler research facility in Barberton, OH for a fully integrated recycle test of the pilot membrane unit with an appropriately sized coal-fired boiler. The field test conducted in fall 2016 demonstrated boiler operation with CO_2-laden air recycled by a membrane system for the first time. Technical results from previous field tests, current small pilot tests at TCM in Norway, and future plans for a 200 tonnes

CO_2/day demonstration system at the Wyoming ITC will be discussed in this presentation.

Cost Analysis of the Uky-Caer Transformative CO2 Capture System

Authors:
Nikolic, H., University of Kentucky
Liu, K., University of Kentucky

U.S. DOE NETL has revised the published reference cases (RCs) used to compare CO_2 capture systems (CCSs). Differences between the RCs will be reviewed, emphasizing supercritical pulverized coal post combustion CO_2 capture, and the University of Kentucky Center for Applied Energy Research (UKy-CAER) CCS cost estimation method will be explained.

UKy-CAER takes a four-pronged approach to CCS. Details about how the synergistic effects of 1) mass transfer intensification; 2) two-stage solvent regeneration; 3) system integration and heat recovery; and 4) the use of an advanced solvent affect the cost of CCS will be presented. The UKy-CAER CCS employs several process intensification techniques for the absorber including partial CO_2 recirculation, temperature profile control, self-adjusting gas-liquid distribution, intercooling and bottom pump-around. Solvent regeneration is conducted in two stages. A split rich feed to the pressurized primary stripper, reduces the steam use and allows for a small column and the secondary air stripper, powered by recovered heat, works to reduce the steam requirement in the primary stripper, with higher rich loading, and provide an extra lean solvent to the absorber. The steam extracted to the CCS reboiler is desuperheated by heating a portion of the boiler feed water, reducing the exergy loss. An advanced solvent with an achievable regeneration energy of 900 BTU/lb CO_2 captured is assumed, with experimental results presented from the UKy-CAER small pilot CCS validating this assumption.

Additional equipment pertaining to the UKy-CAER CCS, being 20% of the bare erect cost, is well worth the extra capital when overall performance is considered. The cost of CO_2 capture with UKy-CAER CCS is $39/tonne (excluding transportation/storage), compared to the RC B12B cost of $58/tonne. Additionally, the equipment cost is lower by 13%, the auxiliary power load is lower by 4.7%, and an additional 26 MW of net power are generated.

Engineering-Scale Demonstration of the Mixed-Salt Process for CO2 Capture: Site Preparation

Authors:
Jayaweera, I.
Jayaweera, P.
Perea, E., SRI International

Engineering-Scale Demonstration of the Mixed-salt Process for CO_2Capture:

Project Start and Site Preparation

SRI Internationalâ€™s (SRIâ€™s) MSP, an advanced ammonia-based CO_2capture technology, is a step-change in the solvent-based, post-combustion capture technology development progression. The MSP uses readily-available, inexpensive ammonium and potassium salt solutions in a novel process flow configuration resulting in improved efficiency, reduced use of heat and electrical energy, reduced ammonia (NH_3) emissions, and a high-pressure CO_2 product. The technology has been tested at multiple stages since 2012; proof-of-concept, bench level, and mini-pilot level in the US and Japan.

SRI is currently planning to demonstrate MSP at engineering-scale (~10 MWe) at Technology Center, Mongstad, Norway (TCM). Under the project funded by the US Department of Energy (DE-FE0031588), the team will use the actual flue gas stream from the residue catalytic cracker (RCC) at TCM, which closely resembles coal-power plant flue gas. The specific objectives of the project include: integrate the MSP to TCM plant for testing under dynamic and long-term, continuous steady-state conditions with a real flue gas stream; operate the MSP with advanced heat integration to improve the process efficiencies and determine the optimal regeneration energy requirement; study the solvent and water management strategies; optimize the engineering scale-operation of MSP to achieve high capture rate, high cyclic CO_2 loading, and produce a high-purity CO_2stream at high pressure; and collect critically important data for a detailed techno-economic analysis and further process advancements to enable the technology to meet the DOEâ€™s \$30/tonne of CO_2 goal by 2030. The project objectives are designed to identify the technology gaps and to determine the cost associated with integrating a MSP-based CO_2 capture system for a >550-MW coal-based power plant.

In this presentation, the MSP technology background and the activities relating to the preparation for the engineering-scale demonstration will be discussed.

Paths to Affordable Deployment of Ccus for a Power Plant Fleet in China

Authors:
Singh, S., Nice America Research, Inc
Hao, P., China Energy
Liu, X., China Energy
Wei, C., NICE
Xu, W., National Institute of Clean-and-Low-Carbon Energy
Ku, A., NICE

The deployment of carbon capture, utilization and sequestration (CCUS) at power plants is currently considered to be too expensive in many parts of the world. In this paper, we show a pathway exists for affordable deployment of CO_2 capture in the Chinese power sector scaling to greater than 100 MM tons per year by 2030. Using a granular analysis of the existing power plant fleet at a large power company accounting for about 10% of total nameplate generation capacity in China, we show that local variations in the cost of CO_2 capture within a power plant fleet result in a discount of up to 30%, relative to the mean values estimated by generalized analyses of reference cases. This paper describes efforts at NICE to identify and demonstrate the affordable paths to CO_2 emissions reductions at a scale of 100 MM tons/year or greater within China Energy Group (CE) operations by 2030. Given that the power sector in China accounts for about 50% of its national CO_2 emissions and about 15% of global anthropogenic CO_2, progress in CCUS at CE can result in meaningful CO_2 emissions reductions while also providing a roadmap for global scale impact.

This paper will cover a techno-economics study of carbon capture across a major portion of the CE coal power generation fleet. Current CE assets include over 150 GW of coal-fired power generation distributed across China. The paper will provide a clearer picture of cost of electricity (COE) and capture costs and avoided costs when the fleet is retrofitted with carbon capture technology. This study will be used to guide efforts to meaningfully and affordably reduce carbon intensity across the entire system. Initial CCUS deployment pathways, using commercially available technology, will be shown. R&D targets that can enable meaningful improvements in affordability will also be presented.

Feasibility of a Commercial-Scale Carbon Storage Project in North Dakota

Authors:
Wildgust, N., University of North Dakota
Peck, W. D., University of North Dakota
Hamling, J. A., University of North Dakota
Gorecki, C. D., University of North Dakota

The Energy & Environmental Research Center is investigating the feasibility of storing upwards of 50 million tonnes of CO_2 in central North Dakota over a 25-year operational period. Technical and nontechnical challenges specific to commercial-scale deployment of a CO_2 storage project are being addresssed, with a long-term goal of developing a permitted geologic storage opportunity should a business case for CO_2 storage emerge. Two geologic storage areas located adjacent to separate lignite-fired energy generation facilities are being evaluated. Analysis of the newly acquired core shows sandstone porosity in the 25% to 35% range with corresponding permerabilities of approximately 1000 mD, suggesting that the Broom Creek Formation is an excellent target for CO_2storage. 3-D geocellular models provide the foundation for dynamic simulation efforts to predict potential CO_2 and pressure plume extents at both storage areas based on the premise of a 2-million-tonne-a-year injection rate. Understanding the potential extent of CO_2 provides guidance on pore space leasing requirements; monitoring, verification, and accounting activities; and development of business case scenarios. Modeling results indicate over 200 Mt of CO_2 storage resource in each of the two areas under investigation. Support for feasibility is bolstered by an existing North Dakota CO_2 pipeline, long-term liability laws for CO_2 storage, pore space ownership laws, and primacy to administer the Environmental Protection Agency (EPA) Class VI underground injection control (UIC) rules. To support future phases of this project, a permitting plan incorporating local, state, and federal regulations is being developed with guidance from officials who oversee North Dakotaâ€™s primacy of EPA Class VI UIC wells. Project partners include DOE, the North Dakota Industrial Commissionâ€™s Lignite Research Council, ALLETE Clean Energy, Basin Electric Power Cooperative, Minnkota Power Company, BNI Energy, North American Coal, Prairie Public Television, Schlumberger Carbon Services, and Computer Modelling Group, Ltd.

Optimizing Carbon Utilization and Sequestration Efficacies in Sandstone Oil Reservoirs Using an Integrated Machine-Learning Workflow

Authors:
Sun, Q., New Mexico Institute of Mining and Technology
You, J., New Mexico Institute of Mining and Technology
Ampomah, W., New Mexico Institute of Mining and Technology
Kutsienyo, E. J., New Mexico Institute of Mining and Technology
Balch, R., New Mexico Institute of Mining and Technology
Cather, M., New Mexico Institute of Mining and Technology
In this study, a robust optimization framework incorporated with multiple machine learning algorithms is structured to co-optimize a CO_2 enhanced oil recovery (EOR) project considering carbon sequestration and hydrocarbon extraction efficacies. A field-scale numerical simulation model is employed to simulate a water alternative CO2 injection (WAG) project in Pennsylvanian Upper Morrow sandstone reservoir in the Farnsworth Unit (FWU), Texas.

A compositional model that couples numerous field data (geological, geophysical and engineering data) is established to serve for the optimization scheme. The model was calibrated based upon the field data to generate a successful history-matched model. The overall objective function for the optimization study is assigned by combining parameters such as oil recovery factor, CO_2 storage and net present value. In this work, neural network models are trained and employed as proxies of high-fidelity numerical models to reduce the extensive computational overheads involved in the optimization process. Simulation data generated from high-fidelity model were utilized to train the proxies via iterative calibration runs. Once an accepted proxy was realized, were it would be applied in evolutionary and machine learning optimization algorithms to find the optimal solution for the pre-defined objective function. Engineering design parameters to be optimized include well bottomhole operational pressure, water and gas injection rate, WAG cycle durations, etc. The robust optimization workflow developed in this work is competent to find the field development strategies that maximize both hydrocarbon production and CO_2 sequestration in FWU. It takes advantage of the fast computational speed of the neural network models to assist numerical simulations for field-scale CO_2-EOR projects. Moreover, the observations and lessons earned in this work would provide significant insights to similar projects throughout the world.

Quantifying CO2 Injection Performance in Depleted Reservoirs for Operational Optimization in CCUS Projects

Authors:
Ravi Ganesh, P., Battelle Memorial Institute
Mishra, S., Battelle Memorial Institute
Gupta, N., Battelle

Carbon dioxide (CO_2) injection into depleted oil reservoirs routinely implement subsurface models to track and understand multiphase system dynamics. This study uses full-physics compositional simulations of a depleted reservoir with a simplified radial geometry to numerically investigate the efficiency of horizontal wells for our program partners to plan field workover activities for similar reservoirs in the MRCSP project in northern Michigan, USA. The model consists of a single vertical well in the center which acts as a producer during an initial primary recovery phase and is then converted into a CO_2 injector during a subsequent CO_2 injection period with predetermined injection-falloff periods. The simplified configuration enabled numerical investigation of different desirable basic well configurations such as (a) horizontal well versus vertical well with sensitivities to horizontal well length and (b) vertical well with lateral sections of different well lengths for reservoirs of varying permeability.

The injectivity index is a commonly used concept in petroleum reservoir engineering to evaluate the wellâ €™s capability for injecting fluids into a porous and permeable formation. Our earlier work established that pressure and rate data can be successfully integrated using this metric to assess and compare performance of different wellbore-reef complexes [Mishra et al., 2017 and 2018]. Numerical evaluation of improvement in injectivity index in reservoirs with different absolute permeabilities is performed using sensitivity analyses of above-mentioned injector configurations. Horizontal well sections are found to be more efficient as expected with higher injectivity index corresponding to longer horizontal well cases. Cross-plots reveal clear trends that correlate the injectivity index with the intrinsic reservoir permeability and the horizontal well length.

This work establishes a reliable quantitative performance comparison between different injection well configurations using a simple yet effective injectivity index metric. Such simplified models can be valuable resources for cost-benefit analysis by operators during preliminary project optimization.

The Development of CO2 Plume in CO2 Sequestration in the Aquifer

Authors:
Long, Y., Missouri University of Science and Technology
Fu, H., University of North Dakota
Wang, S., University of North Dakota
Ling, K., University of North Dakota
Yu, P., Beibu Gulf University
Wang, Y., University of North Dakota

Geological carbon sequestration through injecting large-scale CO_2 into the deep saline aquifers represents long-term storage of CO_2. In the CO_2 sequestration process, the injected CO_2 is displacing water from the injection point and is expected to remain in the reservoir. Due to the nature of one phase displacing another phase in porous media, it is noted that different water saturation exists in the CO_2 plume during the displacement. The distribution of water saturations behind CO_2 front during the displacement is a function of fluid and rock properties, fluid-rock interaction, and injection operation. In this study, these factors are considered when developing new models to predict CO_2 plume evolution during injection. Mass conservation, multiphase flow, and equations of state are applied in the derivation of the models, which guarantees a rigorous approach in the investigation.

The modeling results indicate that CO_2 does not displace water entirely away from the plume. The shape of the CO_2 front is controlled by the relative permeability of two phases and capillary pressure. Water saturation profile from CO_2 injecting point to the displacement front shows that water saturation behind the CO_2 front increases outwardly, and the change in saturation is non-linear. The injection rate impacts the sharpness of the CO_2 front, thus leads to different gas plume sizes for the same injection volume. The outward movement of the CO_2 front decelerates as injection time goes on.

The research illustrates that injection experiences two stages: transient and steady-state, in which the displacement behavior and the development of gas plume vary. Although the duration of the transient stage is dictated by the size of the aquifer and is relatively short compared with steady-state stage, its influence on the development of CO_2 plume cannot be neglected when selecting gas compressor horsepower and determining injection rate.

Wyoming's CarbonSAFE Phase II Project at Basin Electric's Dry Fork Station

Authors:
Carpenter, S. M., University of Wyoming
Coddington, K., University of Wyoming
McLaughlin, J. F., University of Wyoming
Quillinan, S., University of Wyoming

Building on the prefeasibility work completed by the University of Wyomingâ€™s Project Team during the first phase of the Wyoming CarbonSAFE project (DE-FE0029375), the Department of Energy has funded the Feasibility Phase II (DE-FE0031624) to be conducted at Basin Electricâ€™s Dry Fork Power Station (DFS), near Gillette Wyoming.

The project team consists of a highly qualified team of companies and subject matter experts. The project team is organized into three topical areas; Site Characterization, Community Outreach and Economics, and Future Site Plans. The project team consists of the site host, Basin Electricâ€™s DFS in Gillette. The internal University of Wyoming team members include the UW Center for Economic Geology Research, UW College or Engineering and Applied Science, UW College of Law, UW College of Business, and the Enhanced Oil Recovery Institute. External project team members include; Advanced Resources International, Carbon Geocycle, KKR, Energy and Environmental Research Center at UND, and Schlumberger.

Phase II Goals and Objectives include three main types of objectives; technical, non-technical, and operational. Specific technical objectives include evaluating storage complex feasibility; identifying risks that have the potential to prevent the storage complex from serving as a commercial storage site; developing commercial-scale approaches to monitoring, verification, and accounting (MVA), and generating analyses and approaches to support UIC Class VI well permitting. The non-technical objectives include refining and updating the projectâ€™s commercial scale economic model, regulatory analysis, and CO_2 management strategy surrounding DFS; implementing the DFSpublic outreach plan; and preparing a Phase III site characterization plan. Finally, the two operational goals include drilling a successful stratigraphic test well and acquiring new 3D seismic data.

This presentation will discuss the transdisciplinary team and approach to UWâ€™s Phase II CarbonSAFE Project that endeavors to advance commercial scale CCUS, incorporating the potential for beneficial reuse of CO_2, saline storage of CO_2, and the possibility of CO_2 storage incidental to CO_2-EOR.

Direct Air Capture with CO2 Storage - a backstop technology for achieving net-zero

Authors:
Gibbins, J., UK CCS Research Centre

Technologies that can use machinery rather than natural systems to permanently remove CO_2 from air are an essential â€˜backstopâ€™ for achieving net zero emissions. Whatever the cost of this direct air carbon capture and storage (DACCS) turns out to be, this value effectively defines the real cost of emitting CO_2 to the atmosphere. So bringing this cost down as fast as possible could be the deciding factor in getting serious global commitment to the necessary finite national, and hence global, CO_2 emission budgets.

So far most non-government venture capitalist and similar support has been for direct air capture (DAC) to produce CO_2 to make products such as synthetic fuels for boutique markets, as we are seeing now. This has been very valuable to get things moving, but in any serious emission reduction programme large amounts of storage of CO_2 from the air as an offset for other emissions will be required. Some examples will be discussed.

DACCS is not â€˜geoengineeringâ€™, just the equivalent of picking up litter; the key is to get the litterer to pay, not pass the cost on to future generations. If DACCS can be driven through the same type of cost reduction process as solar power, then no-one can claim that net-zero emissions is impossible for any process.

For hard-to-avoid emissions, DACCS is likely to be faster and cheaper to implement than radical changes in the way we do things. For example, if DACCS costs can be brought down to even Â£200 per tonne of CO_2 permanently removed from the atmosphere (and developers hope to do much better than this) then pump prices for the ultimate cut in emissions to net zero fossil petrol and diesel would need to go up by only about 50p/litre. DACCS could also be the best way to offset infrequent CO_2 emissions from natural gas central heating systems used to top up hybrid heat pumps.

Techno-Economic Modeling of Dual-Purpose LNG CO2 Shipping

Authors:
De Leon, R., University of Houston
Datta, A., University of Houston
Krishnamoorti, R., University of Houston
Source to sink matching is a critical determinant for the feasibility of Carbon Capture, Utilization, and Sequestration (CCUS) projects as transportation costs scale up with distance. Typically, sources of CO_2 like chemical and power plants are not in close proximity to potential use and storage (sinks), thereby rendering CCUS projects infeasible. The use of pipelines to transport CO_2 has added excessive capital costs and the threat of sunk capital costs to existing projects. Utilizing the explosive growth of the LNG/LPG shipping industry can substantially lower the costs and de-risk transportation of CO_2 for CCUS. In this scheme, importers of natural gas capture CO_2 and transport it via specially designed dual-purpose vessels to exporters that have aging fields offshore or near to shore for utilization through enhance oil recovery. In this work, we analyze the techno-economic feasibility of this proposition to present the incentives that dual-purpose ship transport can provide- transportation over long distances at a fraction of prevalent costs, elimination of operating costs for an empty vessel on its return journey, the flexibility to route CO_2 where itâ €™s needed, and the economic benefit of sequestration. The results demonstrate that transportation costs are not a bottleneck in the value chain and can be reduced by an order of magnitude from \$16-\$30 to \$2-\$3 per ton of CO_2. We discuss specific source to sink matching for importers that are likely to rely heavily on fossil fuels to meet their energy requirements. This provides an opportunity to reform current economics and policies around carbon management for climate change mitigation with a three-way payout mechanism to the capture, charter and sequestration stakeholders.

The Economics of Accelerated Carbonation Routes Using Captured CO2

Authors:
Strunge, T., IASS Potsdam
Naims, H., Institute for Advanced Sustainability Studies e.V. (IASS)

A major hurdle for implementing CCU technologies is often their economic viability as well as the social acceptance for using such technologies [1]. Therefore, assessments regarding the economic and social impacts of CCU technologies are needed.

Being among the biggest emitters of anthropogenic CO_2, the cement industry requires affordable pathways towards a sustainable future [2]. A concept developed in this field is the so called â€˜accelerated carbonationâ€™ process. Hereby, CO_2 is reacted with activated minerals to form carbonates, which could potentially be used for multiple purposes, such as fillers, cement additives or for land reclamation projects [3,4]. Some policy advice reports use the accelerated carbonation process as a positive example for the utilization of CO_2 as a feedstock, because unlike most other CCU concepts, the carbonation reaction is energetically favorable [5]. Although the concept is not new, accelerated carbonation routes lack detailed and comparable economic assessments in literature [3]. In this contribution, economic assessments of several carbonation routes will be presented, uncovering possible advantages of certain routes towards an economically viable implementation. Moreover, the evaluation of the circumstances under which these novel technologies become economically feasible as well as the analysis of key factors to promote economic feasibility will be investigated.

Understanding the economics of accelerated carbonation routes is essential for their further development as well as deployment and can help decision makers to derive sustainability strategies.

References

[1] A. Zimmermann, M. Kant, T. Strunge, E. Tzimas, W. Leitner, W. Arlt, P. Styring, K. Arning, M. Ziefle, R. Meys, **2017**.

[2] A. Favier, C. De Wolf, K. Scrivener, G. Habert, ETH Zurich, **2018**.

[3] A. Sanna, M. Uibu, G. Caramanna, R. Kuusik, M. M. Maroto-Valer, *Chem Soc Rev* **2014**, *43*, 8049-8080.

[4] A. Sanna, M. R. Hall, M. Maroto-Valer, *Energy & Environmental Science* **2012**,*5*, 7781.

[5] WWF Deutschland, **2018**.

Toward Improved Guidelines for Cost Evaluation of CO2 Capture Technologies

Authors:
Fout, T. E., U.S. DOE National Energy Technology Laboratory
Roussanaly, S., SINTEF Energy Research, Norway
Rubin, E. S., Carnegie Mellon University
Van Der Spek, M., ETH Zurich
Berghout, N., International Energy Agency
Booras, G., Electric Power Research Institute
Garcia, M., IEAGHG
Gardarsdottir, S., SINTEF Energy Research
Ho, M., University of Sydney
Matuszewski, M. S., AristoSys, LLC, Contractor to National Energy Technology Laboratory
Nazir, S. M., NTNU
Ramirez, A., Delft University of Technology

One of the key barriers to wide-scale application of CCS is its cost. In practice, there are many challenges in establishing reliable cost estimates for CCS technologies. Building on the first set of cost guidelines established under the IEAGHG cost network, we are working on three areas where further cost guidelines are needed.

The first area that this group aims to tackle is the establishment of improved guidelines for cost evaluation of CO_2 capture technologies that are not yet commercial, and the evolution costs beyond demonstration projects. While emerging technologies may have the potential to reduce costs, most techno-economic assessments are performed considering them as mature, introducing a significant bias in their comparison. Our activity in this area is expected to provide a better understanding of cost evolutions from early stages of process development to the first large-scale plant, and then onwards to the Nth-of-a-kind plant for the successful technologies.

The second area to be tackled is the need for transparency, data quality and uncertainty evaluations of both the data and models used in CCS cost analysis. In practice, different cost estimates can be obtained for the exact same process, depending on the assessment methodology, the quality of the applied data, and the level of detail considered. The work in this area will emphasise the needs for improved data transparency and uncertainty evaluation, as well as provide guidelines on which methods to use for a certain task.

Finally, with the growing focus and need for CCS from non-power industries, improved guidelines for the evaluation of CCS for non-power industries are also in development. This area addresses key elements for evaluation of CCS from industrial plant such as heat and power supply strategies, retrofitting costs, the impact of scale of captured CO_2 on transport and storage costs, etc.

Direct Air Capture - Advances and Opportunities

Authors:
Cirucci, J. F., Arizona State University
Lackner, K. S., Arizona State University
Wright, A. B., Arizona State University
Page, R., Arizona State University
CO_2 direct air capture technology continues to undergo significant technology, engineering and early commercialization advances. Multiple pilot and demonstration facilities employing different capture technologies have been in operation, and niche commercial facilities are now onstream. Yet skepticism regarding direct air captureâ€™s future is still widespread. Surprisingly, the thermodynamic climb up from 400+ ppm CO_2 concentration does not impose an insurmountable economic barrier. Rather, Direct Air Capture eliminates many location constraints associated with point source capture and provides important supply chain advantages and opportunities. However, there remains key challenges and unique solutions. Here, the state of Direct Air Capture will be reviewed, dissecting the perceived obstacles to technical and economic acceptance, and comparing the different process technologies employed or proposed. A new design and techno-economic analysis for a â€œpassiveâ€ system which does not impose the substantial energy demand associated with forced convection of bulk air will be described.

Incremental Time-Lapse Monitoring of CO2 EOR and Geologic Storage Using a Scalable Automated Semipermeant Seismic Array

Authors:
Livers-Douglas, A., Energy & Environmental Research Center
Burnison, S., University of North Dakota
Barajas-Olalde, C., University of North Dakota
Hamling, J. A., University of North Dakota
Wildgust, N., University of North Dakota
Peck, W. D., University of North Dakota
Gorecki, C. D., University of North Dakota

The Energy & Environmental Research Center conducted proof-of-concept field testing of a scalable automated semipermanent seismic array (SASSA) and trace-based processing and interpretation method that can detect CO_2 saturation changes at discrete locations in a reservoir undergoing CO_2 injection for enhanced oil recovery (EOR) or geologic storage. Using a sparse array of geophones and a stationary source, SASSA is capable of operating remotely and autonomously and has a greatly reduced environmental and operational footprint compared to traditional time-lapse 3-D seismic surveys. The trace-based method allows for rapid processing and interpretation that could ultimately contribute to same-day decision making. The SASSA technique was demonstrated at an active CO_2 EOR project during a 12-month field-test which acquired 41 sets of time-lapse data using 96 three-component geophones and a single remotely operated accelerated weight drop source. The technique detected CO_2 saturation changes that were validated against a conventional time-lapse 2-D line and advanced reservoir simulations. While the test resulted in a proof-of-concept, many of the records were impacted by noise.

A second field demonstration is under way that incorporates lessons learned from the initial field test to improve signal-to-noise ratio. Specifically, the demonstration will incorporate stronger sources, increase the shots-per-acquisition to improve stacking, incorporate strategic placement of geophones to reduce noise, and acquire complementary data sets regarding weather and ground conditions to develop tools for processing noise and to determine optimal conditions for system operation. Two orbital vibrators at offset source locations will operated consecutively during each acquisition as a technique to avoid the ground roll zone and improve coverage. The second field demonstration will provide key information about system performance, cost, designs and processing to advance the next-generation SASSA monitoring technique toward commercialization.

Integrated Midcontinent Stacked Carbon Storage Hub Phase II: Storage Complex Data Collection Planning

Authors:
Fukai, I., Battelle Memorial Institute
Duguid, A., Battelle Memorial Institute
The Integrated Midcontinent Stacked Carbon Storage Hub (IMSCS-HUB) is part of the Carbon Storage Assurance Facility Enterprise (CarbonSAFE) established by the United States Department of Energy (DOE) National Energy Technology Laboratory (NETL) to support development of an integrated, commercial-scale carbon capture and storage (CCS) complex for operation in the 2025 time frame. The IMSCS-HUB study region consists of a CO_2 source corridor and a stacked storage corridor that spans across Nebraska and Kansas, representing the first large-scale project for the Midcontinent region of the United States. As part of Phase II of the project, the overall goal is to conduct a detailed feasibility assessment of geologic CO_2 storage complexes at potential sites in southwest Nebraska and Kansas.

This presentation will describe the data collection planning effort to identify gaps in geologic, model, and regulatory data that could be addressed via new data acquisition to support Phase II objectives and prepare for future phases of the project. This involves updating existing geologic models with recently available public and proprietary data and conducting a geologic and regulatory data gap analysis for each potential site. Plans for data acquisition and analysis from new characterization wells at each site will also be discussed, including: whole core acquisition and analysis, well tests, fluid sampling and analysis, basic and advanced wireline logging, and seismic data acquisition.

The new data collected as part of Phase II field activities will support the feasibility analysis and initial characterization of the storage complexes at each potential site. This includes data to be used for subsurface characterization and modeling, outreach activities, risk assessment, regulatory analysis, and permit planning to establish commercial development strategies for one or more 50-Mt-scale stacked storage sites in the Midcontinent region.

Packing Free Gas Absorption Columns for Amine Scrubbing of CO2

Authors:
Fursule, I., University of Kentucky
Nikolic, H., University of Kentucky
Liu, K., University of Kentucky

Conventionally for carbon capture the amine solvent is introduced in the absorption column with packing in a counter-current form to facilitate gas-liquid contact to combat the slow kinetics thereby increasing the mass transfer for the gas absorption. But the capital cost of packed bed is a dominating drawback to implement the CO_2 capture to the power industry. The University of Kentucky Center for Applied Energy Research (UKy-CAER) is introducing a packing free approach to the absorption column that makes it possible to alter the current packed absorption column into a continuously rising froth column of the amine solvent using additives where the froth bubbles essentially act as reaction interface. Advantages of this approach include low-pressure drop across the column, higher mass transfer coefficient, zero capital cost of packings, and shorter columns. With the inclusion of frothbed instead of structured packing in the column, the height of the column can be reduced to less than 50% of the original for the same efficiency of gas absorption keeping all the operational parameters same. A comparative parametric study is conducted using two different solvents, 5M 1-Amino-2propanol (new generation solvent), and UKy advanced solvent and three different types of surfactants. Primarily the stability of froth is related to the type and concentration of the surfactant, typically increasing from 0.001 wt% to 0.1 wt%. Froth rheology is studied under the superficial gas velocity between 0.05 m/s to 1 m/s. Results demonstrating the tunability of the physical structure and rheology of froth using the process parameters, surfactant, and solvent properties at different carbon loadings will be presented in detail.

Sensor and Control Technologies for Geologic CO2 Storage Operations - a Doe Carbon Storage Project Review

Authors:
Aljoe, W., U.S.Dept. of Energy
Mahajan, K., National Energy Technology Laboratory
Rodosta, T., National Energy Technology Laboratory
The Carbon Storage Program implemented by the U.S. Department of Energyâ€™s (DOE) Office of Fossil Energy and managed by the National Energy Technology Laboratory (NETL) is helping to develop technologies that safely and permanently store carbon dioxide (CO_2) without adversely impacting natural resources or hindering economic growth. Since its beginning in 1997, the Program has significantly advanced carbon capture and storage (CCS) science and technology; however, key gaps in experience and knowledge must be overcome before CCS technology can be widely implemented by operators in the field. One of these gaps is related to tools and technologies used to monitor CO_2 in deep subsurface geologic storage complexes. Technologies are deployed to determine the location of the CO_2 plume, to provide assurances that the injected CO_2remains in the subsurface, to detect the quality of ground water resources, and to ensure that pressure from the injection operation does not damage subsurface resources or compromise the integrity of the storage complex. Sensors utilized for carbon storage can include, but are not limited to, fiber optic sensor systems, distributed acoustic sensing systems, tools to measure physical attributes of the system (e.g., temperature, pressure, strain, chemical properties), and geophysical methods. Ideally, inputs from multiple sensors would be incorporated into intelligent monitoring systems for controlling carbon storage. DOE-NETL supports a wide variety of projects whose aim is to develop and test innovative technologies for sensing the parameters required for controlling CO_2 injection in subsurface carbon storage operations. This presentation will review the status of CO_2monitoring technologies supported by DOE, focusing on sensors and controls employed in carbon storage. Tools and technologies intended for atmospheric, near-surface, and subsurface monitoring will be discussed.

The Carbon Capture and Storage Readiness Index: Is the World Ready for Carbon Capture and Storage

Authors:
Consoli, C., Global CCS Institute
Havercroft, I., Global CCS Institute
The Carbon Capture and Storage (CCS) Readiness Index (CCS-RI) identifies those nations which are leaders in the creation of an enabling environment for the commercial deployment of CCS, known as CCS Readiness. The progress of CCS deployment is tracked through a series of targeted â€˜indicators' that consider a countryâ€™s:
â€¢ Inherent CCS interest
â€¢ Policy developments
â€¢ Legal and regulatory frameworks
â€¢ Geological CO2 storage development.

The 2018 CCS-RI examined over 50 countries using 70 discrete criteria to create an overall score for each nation, enabling a comparative assessment of countries globally. The results found that only five countries rank in the Indexâ€™s highest category taking significant steps to reduce domestic barriers to CCS. Overwhelmingly, the majority of nations examined are not ready for the wide-scale deployment of CCS. In most instances, low national scores are a result of weaker policy settings, but not always. The outcomes of the CCS-RI will be discussed including identifying those barriers to deployment and discussing those nations that are well-advanced and primed to deploy over the coming decade.

A Study of Ultra-Fast CO2 Desorption Kinetics for a Microwave-Enhanced Regeneration Process

Authors:
Shi, F., National Energy Technology Laboratory
Gray, M. L., U.S. Department of Energy, National Energy Technology Laboratory
Ji, T., National Energy Tech. Lab.
Soong, Y., National Energy Technology Laboratory

Regeneration of CO_2-rich solvents plays a significant role in carbon capture technology. In a traditional aqueous amine-based post-combustion carbon capture process, thermal regeneration cycles are usually conducted using superheated steam. It takes several hours to swing an aqueous amine solution over a temperature range. Furthermore, a significant amount of make-up water is required for each steam regeneration step. This implies an inherently slow process and a huge energy penalty (as a large amount of solvent is water which also needs to be heated or cooled down), when the conventional temperature swing absorption process is used.

In this work, a microwave (MW)-enhanced regeneration process was able to provide ultra-fast CO_2 desorption from an aqueous amine solution and a non-aqueous solution. Due to their polarization properties, amine-CO_2 pairs in the solution are strong MW energy absorbers. With localized MW heating on amine-CO_2 pair molecules, absorbed CO_2 can immediately dissociate from amine groups even when the bulk temperature is as low as at 70 °C. Based on the kinetics of desorption, the activation energy of CO_2 desorption under MW conditions was significantly reduced compared to traditional thermal regeneration. The mechanical study implied that hot-spots on the amine groups may trigger the desorption reaction and then alter reaction pathways. With this MW-enhanced regeneration process, ultra-fast desorption kinetics and possible low temperature regeneration are able to be achieved. As a consequence, the huge capital investment due to large equipment size, the large energy penalty required for solvent regeneration, and the significant thermal-degradation of amine in the traditional regeneration process may be avoided.

Controlled Synthesis of MgO with Diverse Basic Sites and Its CO2 Capture

Authors:
Gao, W., Beijing Forestry University
Wang, Q., Beijing Forestry University

Novel sodium dodecyl sulfate (SDS)â'assisted mesoporous MgO adsorbents with diverse basic sites were prepared hydrothermally via a urea hydrolysis synthesis method for intermediate CO_2 capture. The control of synthesis parameters and the introduction of SDS surfactant greatly improved the capture performance. With elevated hydrolysis temperatures, the phase transition process made great contributions to the morphological changes of the precursor architectures. In situ DRIFTS analysis demonstrated that various carbonate surface species including bicarbonate, bidentate, and unidentate carbonates were formed on the obtained MgO during interaction with CO_2. Furthermore, the main component of adsorbed CO_2 surface species swings from bicarbonate to bidentate and unidentate carbonates with increase of adsorption temperature. The highest CO_2 uptake of 1.22~1.99 mmol gâ'1 was attained for MgO sample at the lowest calcination temperature in a wide temperature range of 60~300 °C. High specific surface area (372.0 m^2 gâ'1), large pore volume (0.38 cm^3 gâ'1) as well as diverse basic sites of the synthesized MgO make it an eligible candidate for CO_2 capture from certain sorption enhanced hydrogen production processes, with a nearly 20â'fold enhancement of the commercialized light MgO. Additionally, the results of CO_2uptake studied under diluted and wet (H_2O containing) CO_2 conditions for the asâ'prepared MgO adsorbent also suggested good prospect in practical applications.

Experimental Evaluation of Aerosols Formation and Control during CO2 Absorption into Monoethanolamine Solution

Authors:
Salih, H., University of Illinois at Urbana-Champaign
Nelson, P., University of Illinois at Urbana-Champaign
Ye, Q., University of Illinois at Urbana-Champaign
Oki, A., University of Illinois at Urbana-Champaign
Lu, Y., University of Illinois at Urbana-Champaign

Solvent loss from air emissions in form of vapor and aerosols is one of the main challenges for aqueous amine-based CO_2 absorption processes due to its negative impacts on the environment and human health as well as the operating cost. The main objective of this research is to evaluate the emissions of solvent from the absorber and the effectiveness of a water wash section to reduce these emissions. In this research, laboratory experiments were conducted to study solvent emissions, in form vapor and aerosols, from a CO_2 absorption column using 5 M MEA solution and through a water wash section. The vapor concentrations were measured using a membrane filtration device and FTIR instrument. While the particle size distribution and the concentrations of aerosols were obtained using two particle size spectrometers. In general, the aerosols formation was found to be controlled by solvent properties temperature gradient and the quality of the inlet gas specially the concentration and types of the entrapped particles. Results of our baseline experiments revealed that in the absence of CO_2 absorption, a majority of solvent emissions were in vapor form. Solvent emissions generally increased with decreasing feed CO_2 loading. The removal of MEA vapor by the water wash section was not significant (~10%), while up to 70% of the aerosol emissions were removed by the water wash. The size of aerosols increased from 52 nm in the feed gas to up to 257 nm in the effluent gas from the absorber due to particle growth and aggregation. By contrast, in the water wash section, the size of aerosols generally decreased because of the removal of larger particles.

Hydrothermal Processing of Food Waste for Resource Recovery

Authors:
Jena, U., New Mexico State University
Bayat, H., New Mexico State University

With the growing global population, problems like providing clean water, food, and energy increase dramatically as resources are depleted faster. The development of new energy sources has become particularly important from the perspective of energy security and environmental protection. According to the Food and Agriculture Organization of the United Nations, one-third of global food production is wasted in a world where 900 million people are suffering from hunger. High moisture food waste could be converted into useful products (water, fertilizer, liquid fuels) in a wet process referred to as hydrothermal liquefaction (HTL). HTL is an alternative technology to conventional waste treatment due to great ability to process high moisture content biomass waste without prior dewatering which produces an energy-dense liquid fuel precursor, called â€˜bio-crude oilâ€™. The purpose of our research is experimentally investigating the optimized HTL conditions (operating temperature, reaction time and etc.) to improve bio-crude oil yield and its quality from food waste. HTL product characterization includes higher heating value, elemental composition, and chemical constituents. Our preliminary results show that bio-crude oil generated from HTL process has an energy content of 30-40 MJ/kg and HTL of food waste can be effective for organic waste management and bioenergy production. This presentation will provide an overview of HTL of food waste and its co-product utilization.

Investigation of Coal Wettability for the CO2 Sequestration and ECBM Applications: A Review

Authors:
Ibrahim, A., Texas A&M University

The efficiency of the enhanced-coalbed-methane (ECBM) and CO_2 sequestration process depends on the wettability behavior of the coal-water-CO_2 system. It depends on the CO_2 diffusion rate from the cleat network, through the micro-cleats, to the surface of the coal matrix. If the coal is hydrophobic, the gas will fill the smaller pores, which leads to faster diffusion of CO_2 to the coal surface (diffusion coefficient of CO_2 = 1.710^{-7} m^2/s at 100 bar and 300 K). The diffusion coefficient decreases to 210^{-9} m^2/s at the hydrophilic system.

This paper interoperates the published experimental outcomes of coal wettability at varies range of pressure, temperature, and gas composition. Wettability studies usually used the contact angle measurements as a simple method to characterize the coal. Adsorption isotherm, Zeta potential measurements were used to interoperate the contact angle results and understand the wettability behavior. Recently, Fourier Transform Infrared Spectrometer (FT-IR) technique was employed to investigate the functional chemical groups of the oxidized coals.

Coal has different ranks with different oxygen-containing polar groups such as carboxyl (COOH), hydroxyl (OH), and methoxyl (OCH_3) groups. Coal becomes more water-wet with decreasing rank, carbon content and with increasing oxygen-containing groups. Regardless of the coal rank, the coal becomes more CO_2-wet with increasing pressure up to the critical gas conditions where the coal wettability slightly increase. As the water salt concentration increases, it compresses and destabilizes the double-layer surrounding the coal surface, causing a reduction in the absolute zeta potential value and making the coal surface more hydrophobic. Temperature has a negative effect on CO_2 wettability to coal and the coal becomes more hydrophilic with increasing the temperature. EOS can be used to predict the contact angle behavior as a function of the interfacial tension forces. In oxidized coal surfaces, FTIR technique is more sensitive than contact angle measurements.

The Carbonation Kinetics of Alkaline Solids at Conditions Relevant to Direct Flue Gas Utilization

Authors:
Falzone, G., UCLA
Mehdipour, I., Laboratory for the Chemistry of Construction Materials (LC2), UCLA
La Plante, E. C., University of California, Los Angeles
Sant, G., University of California, Los Angeles
Alkaline solids (e.g., $Ca(OH)_2$ or fly ash) can be exploited as cement-alternative binders that also remove CO_2 from industrial flue gases (e.g., from coal/natural gas-fired power plants) via carbonation reactions. When these reactants are mixed with water and aggregates to form a concrete mixture, carbonation causes*in situ* precipitation of carbonate minerals, such as calcium carbonate ($CaCO_3$), that bind particles together and increase the concreteâ€™s strength. The design and optimization of such carbonation processes for the production of concrete components requires detailed experimental datasets and analyses describing the reaction kinetics of the alkaline solids, and a method for scaling up kinetics data to the component level. Such data have not been rigorously evaluated within a parameter space relevant to the direct utilization of flue gases, i.e., temperature (T), relative humidity (RH), and CO_2 partial pressure (p_{CO2}), without substantial pre-processing of the gas stream. To this end, we (1) investigate the relationships between the carbonation kinetics of alkaline solid reactants and their theoretically predicted carbonation potentials and (2) experimentally assess the sensitivity of carbonation kinetics to RH, T, and p_{CO2}. Provided that a critical RH is exceeded, a reactantâ€™s carbonation sensitivity relative to its theoretical carbonation potential can be described by a reduction factor that is correlated to its rate of calcium release, suggesting that carbonation kinetics in this regime are dissolution-limited. The carbonation kinetics of mixtures of powder reactants are found to be strongly predicted by a linear rule of mixtures, indicating negligible influences of concurrent pozzolanic reactions during the processing period. These findings provide critical inputs that inform the design of reactant mixture formulations and carbonation processes in relation to the properties of the constituent reactants. Future research is directed towards extending this approach for the scalable production of carbonated concrete components by direct flue gas utilization.

Incentivizing Carbon Utilization Research

Authors:
Chavez, A., Northern Kentucky University
Policies to incentivize carbon capture and utilization need to be developed. Governments have used a number of approaches to stimulate innovation. These have primarily consisted of patents, grants, prizes, and tax incentives. Each method has particular strengths and weaknesses.

Patents, unlike the other incentives, impose the cost of new inventions upon the subsequent purchasers of those products rather than on the governmentâ€™s budget. But they are typically preferred where the value of inventions is uncertain, since they reward inventors through subsequent market activity.

Grants provide resources to innovators while they are still in the development stage, and this can be particularly helpful to researchers with limited funds. Because grants are not contingent upon the success of the invention process, however, they do not award achievement, just effort.

Prizes are usually larger than direct subsidies, but they are, of course, less certain. This uncertainty can limit their usefulness for more risk averse, or less well-financed, innovators. Moreover, the government may not be able to calculate accurately the true value of the invention.

Tax credits can provide fairly immediate benefits, but their value requires that the innovators have current income against which to apply the credits. Because the value of tax credits derives from the amounts invested in the invention process, credits may work best where the ultimate value of an invention is hard to determine.

This review of these incentives suggests a number of considerations that inform the best incentive to fit a particular invention process. Considerations include the costs of each policy, the parties who will bear those costs, the circumstances in which each policy is most effective in stimulating innovation, and the timing of the benefit for the recipients.

The range of considerations suggests that governments can use different policies to best incentivize different research goals.

International Standardization As a Tool for Global Commercial Deployment of Ccus

Authors:
Ombudstvedt, I., IOM Law
Jaroy, A. G., IOM Law
Gran, M. E., IOM Law

The overall aim of international standardization is to facilitate the exchange of goods and services through the elimination of technical barriers to trade. ISO/TC265 is a dedicated Technical Committee (â€œTCâ€) to the standardization of CCS and CO_2-EOR. The scope of work is *â€œ[s]tandardisation of design, construction, operation, environmental planning and management, risk management, quantification, monitoring and verification, and related activities(..)â€, covering the whole value chain of CCS.*

Since the start of ISO/TC265 in 2011, the TC has published eight technical reports and standards. The latest addition is standard *â€œCarbon dioxide capture, transportation and geological storage -- Carbon dioxide storage using enhanced oil recovery (CO_2-EOR)â€*, which was published 31 January 2019.

For deployment at a global commercial level, CCS and CO_2-EOR are still left with regulatory gaps, which the standards may help overcome. One example of such a gap is how to properly quantify and verify the amounts of CO_2 being safely stored long-term in order to be eligible for allowances under emission trading schemes around the world. The ISO/TC265 documents aim at closing this gap, enabling e.g. the replacement of natural CO_2 with anthropogenic CO_2 for EOR purposes.

About 20 countries participate to the ISO/TC265, resulting in e.g. reduced costs, dissemination of knowledge, reduction and allocation of risk and liability, the closure of regulatory gaps and enabling of incentives for CCUS worldwide. Some countries end up implementing the standards into their regulatory framework. Norway has as an example, included a reference to the standards on transport and geological storage into their ongoing full-scale demonstration project, as one of the final pieces to the regulatory puzzle. Mexico, as another example, is participating to the ISO/TC265 process in order to learn and use the standards as potential building blocks in their regulatory framework currently under development.

Pathways for a Political Consensus Policy Framework to Rapidly Scale the Ccus Industry for Atmospheric Carbon Dioxide Removal

Authors:
Adams, H., Common Capital Pty Ltd

The rapid growth of CCUS is crucial both to aid the transition to a low emission energy sector and to drive the removal of existing atmospheric and ocean carbon dioxide. UN forecasts conservatively require between 100-1,000 billion tonnes of carbon dioxide removal (CDR) to avoid 4.5 degrees of global warming. With the right policy settings, this represents a multi trillion-dollar opportunity for CCUS applications in industries like cement, minerals, forestry, aquaculture, agriculture, chemicals and plastics. However, to meet the pace and scale of RD&D investment the CCUS industry must avoid replicating the problems and deeply polarised politics of climate change and emission reduction policy.

The application of CCUS for CDR risks the same deadlock as climate change and emission reduction policy, without learning lessons from previous policy failures. The paper presents the findings of a research project into options for a policy framework to rapidly scale CCUS and CDR, which can achieve political consensus.

We draw on original structured interviews on CCUS and CDR with a broad cross-section of actors. These include conservative, centrist and liberal legislators; senior bureaucrats; industry executives and lobby groups; institutional and private investors; policy think tanks; consumer groups; environmental organisations; and labor organisations.

We explore the politically viable options to address issues such as: how to fund the scale of CCUS R&D and infrastructure required; how to frame and communicate CCUS/CDR as a policy issue and opportunity; and the role of communications and coalition building to ensure effective technologies arenâ€™t hindered by lack of political or public support. This work builds on existing research into the short comings of climate change policy and political communication.

We consider the implications of these findings for CCUS policy and industry in fossil fuel-rich economies like Australia and the United States.

Testing the ISO Carbon Dioxide Geological Storage International Standard (ISO 27914) Against a U.S. CO2 Storage Site Characterization Program

Authors:
Riestenberg, D., Advanced Resources International, Inc
Zupancic, B., CSA Group
Koperna, G. J. Jr., Advanced Resources International, Inc.
Kemper County in southern Mississippi, USA is home to â€œProject ECO_2S,â€ also known as â€œEstablishing an Early CO_2 Storage Complex in Kemper County Mississippi.â€ Project ECO_2S is one of 16 Carbon-Safe program projects (Phase II) and is a commercial-scale CO_2 storage complex, with financial and technical support from Mississippi Power Company, a Southern Company Affiliate. Storage capacity at the site is on the order of one gigatonne. This effort will drive advancements in understanding of CO_2 storage business arrangements, processes and technologies, and potentially result in accelerating storage of industrial CO_2.

Project ECO_2S, along with other pilot CCS projects, provide significant experiential results which further the feasibility for commercial CCS implementation. Other important work on this front includes the recently published standard for Carbon Dioxide capture, transportation and geological storage standard (ISO27914:2018) developed by the International Organization for Standardization (ISO). This work contains internationally recognized requirements for the permanent geologic storage of CO_2. The standard, which was developed by a group of technical experts from around the world, including the United States, was developed with the intention of promoting an increase in CO_2 storage through globally harmonized practices and technologies.

The overarching objective of this study is to present a comparison of Project ECO_2S to the recently developed ISO27914 standard for geological storage. There is sufficient subject matter overlap between ISO27914 and Project ECO2S to generate useful insights that the Project Team and other proponents of Project ECO2S can use to evaluate current activity and incorporate in to planning for future phases of the CarbonSafe program. The presentation is also intended to serve as an outreach and education piece for the broader CCS community, both domestic and international, who can benefit from learning not only about Project ECO_2S, but the work of ISO Technical Committee 265.

What Parasitic Load? a New Paradigm for Ccus

Authors:
Peck, W. D., University of North Dakota
Hamling, J. A., University of North Dakota
Wildgust, N., University of North Dakota
Gorecki, C. D., University of North Dakota
In the traditional scenario for postcombustion carbon capture from a coal-fired power plant, the addition of a CO_2 capture facility by the plant operator/owner incurs an additional power demand to run the capture facility. This internal draw of power means less electricity is available for sales to customers or the market (revenue). Sale of the captured CO_2 to the oil field will recoup some of the capture costs. Potential savings realized from the recently established 45Q tax credit may offset an additional portion of the capture cost if the power plant operator has sufficient tax burden to capitalize on the full value of the tax credits. An alternate scenario that may be applicable to the coal-fired power plants in North Dakota involves efficiency improvements, smart heat utilization, a means of maintaining baseload generation, and the introduction of a third-party carbon capture entity. The inclusion of this externally/jointly funded third-party entity changes the traditional paradigm. In this scenario, a portion of electricity sales to the grid is replaced by a combination of steam and electricity sales to the capture entity. In other words, the capture entity has become a new customer: large, close, and long-term. The net effect to the power station is essentially zero lost sales. The capture entity will then sell captured CO_2 to an oil field for EOR. In addition, the capture entity can be structured to create a sufficient taxable appetite that can monetize tax incentives available through the 45Q Program. Several studies looking at how this new business scenario could link and benefit North Dakota coal-fired power generation and oil and gas production are under way. This alternative structure and the associated business case could ensure that every entity along the path is a â€œwinner.â
€

CO2 Utilization in Precast Concrete

Authors:
Amr, I., Saudi Aramco
Fadhel, B., Saudi Aramco
Park, S., Korea Advanced Institute of Science and Technology
Hunaidy, A., Saudi Aramco
Bamagain, R., Saudi Aramco
Production of cement is a major source of industrial CO_2 emissions. In recent years, numerous studies have focused on various means of reducing the CO_2footprint of cement-based materials. The present study is related to accelerated carbonation curing technology as a direct mean of reducing the CO_2 footprint by simultaneously utilizing and sequestering it in cementitious products during the curing stage to produce concrete products with enhanced properties and high CO_2 uptake.

Steam curing of cement is commonly used method in manufacturing precast concrete to achieve high early strength. Meanwhile, accelerated carbonation curing produces concrete products with enhanced properties such as sulfate and acid resistance as well as mechanical strength.

In this study, we found that the simultaneous CO_2 and steam curing of precast cement mixture had ~20wt.% CO_2 uptake in cementitious materials, which is almost 5 times higher than the uptake of conventional industrially practiced curing under only steam. Furthermore, the proposed method resulted in 32% increase in the early compressive strength, and that the targeted mechanical strength under the proposed method was attained within 7 days, in comparison to 28 days under the conventional industrial curing practice.

Coupled Techno-Economic Analysis with Life Cycle Assessment for Ccus Evaluation Techniques

Authors:
Al Hunaidy, A., Saudi Aramco

Abstract:

An increase in the human activities since the industrial revolution has led to a number of challenges related to climate change. Researchers have attributed recent climate change to the increasing of greenhouse gas emission concentration in which CO_2 accounts for at least 75% of its content. Many technical routes have been suggested and studied to reduce, mitigate, avoid or utilize CO_2. Among these solutions, utilization of CO_2 through CCUS (carbon capture, utilization and sequestration) seems to be attracting the attention of large organizations recently since it builds complete overall CO_2 management strategy. One of the promising CO_2 utilization strategies is the CO_2 capture and conversion to valuable products that can contribute to the company/industry profit margin. Although many CO_2 capture and conversion technologies have been suggested in literature, in the absence of any standardized tool efforts are costly and time consuming to find out the most optimum CO_2 utilization pathway assuring economic viability. To tackle these challenges, a computer-based analysis tool ArKa-TAC^3 (Aramco KAIST tool for Analysis of CO_2 Capture & Conversion). It is an evaluation tool that assists in developing an effective carbon management strategy through screening of multiple capture and conversion technologies while selecting the best utilization route based on certain objective function. The software takes into account CAPEX and OPEX for a selected carbon management strategy while reporting a full CO_2 life cycle analysis to confirm the CO_2 reduction level. This applicability of this software includes refineries, gas plants, power plants, iron/steel, cement plants and other carbon intensive industries. A CO_2management case study will be presented using Arka-TAC^3 to see how the tool screens different capture and conversion routes to come up with the optimum utilization route based on economics and CO_2 reduction compared to the business as usual.

Keywords: CO_2 conversion, CO_2 technologies, techno-economic analysis

Fine Calcium Carbonate Production By CO2 Mineralization of Industrial Waste Brines

Authors:
Simonetti, D., University of California, Los Angeles
La Plante, E., University of California, Los Angeles
Sant, G., University of California, Los Angeles
Wang, B., University of Wisconsin-Madison

Calcium carbonate production via CO_2 mineralization offers an attractive route for carbon capture and utilization. The process sequesters CO_2 in a stable form of fine precipitated calcium carbonate (PCC), a high-value product with unit price in the range of \$230-\$280/t and a global market projected to reach 99 M tons in 2020. However, two critical barriers hinder CO_2 mineralizationâ€"(i) the need for costly processes such as electrolysis or addition of alkali hydroxides to maintain alkalinity during mineralization, and (ii) the geographic location mismatch between Ca-rich solutions (or brines) and CO_2 generation sources (e.g., power plants), which often renders efficient integration of feedstocks impractical.

To simultaneously address these challenges, a new process is developed to integrate CO_2 mineralization-based calcium carbonate production into existing zero liquid discharge water (ZLD) treatment systems at coal-fired power plants. In this process, Ca-rich streams, such as produced water from oil and gas extraction or carbon storage operations, serve as the Ca-source. An H^+/Na^+ ion-exchange cycle is designed to provision alkalinity during mineralization and regenerate the ion-exchange reagent in solutions with high salinity. Ca-depleted streams are then treated within a centralized ZLD system. The process beneficially utilizes reject streams that are available at substantial quantities in the vicinity of coal power plants within several U.S. regions. In addition, the process utilizes post-desulfurization flue gas from coal-fired power plants without the need for pretreatment. The CO_2 conversion reactions are performed in alkaline streams at ambient pressure, thereby minimizing the energy burden. By beneficiation of industrial waste streams at low energy input, this process offers significant advantages in energy and CO_2 footprint over the traditional high-temperature PCC production. Overall, the new process provides a unique route to integrate CO_2 emissions control and wastewater treatment for coal power plants, while producing a high-value product to offset the economic burden.

CO2 Compression Options for CCUS

Authors:
McKaskle, R., Trimeric Corporation

CO_2 compression is an often overlooked part of CCS. Itâ€™s a lot more than a little black box at the edge of the capture process flow diagram. CO_2 compression from near-atmospheric to pipeline pressure consumes about 10% of the power output from a power generation facility with capture. In fact, the compression parasitic load is about half of the parasitic load for solvent-based capture. Yet, opportunities for reducing compression costs are limited, with some compressors already approaching maximum achievable efficiency. Other challenges can include the need to design for large swings in flow rate and discharge pressure, reconciling equipment selection between small-scale and full-scale projects, and special designs for some processes that capture CO_2 at less than atmospheric pressure.

Discharge pressure and flow must be maintained within relatively narrow ranges for many centrifugal compressors and downstream solutions using low cost multistage centrifugal pumps with great turndown capabilities can be used to take larger pressure and flow swings allowing the more expensive machines to run at their sweet spots. Some capture processes such as membranes result in CO_2capture at less than atmospheric pressure and require creative compressor solutions to reduce the overall cost of CCS.

Many demonstration and first-mover industrial CCS projects are better suited by compressors other than centrifugal compressors, which are likely to be selected for full-scale projects. These can include multistage centrifugal blowers, screw compressors, reciprocating compressors, multistage centrifugal pumps, and others. This paper will provide a broad overview of CO_2 compression options for different suction pressures, flow rates, and discharge pressures and for project scales ranging from small-scale demonstration to full-scale commercial projects. Emerging CO_2 compression technologies will also be discussed. A generalized comparison of refrigeration-based liquefaction followed by pumping vs. straight compression will also be presented.

Exploring the Gulf of Mexico Sub-Sea Floor As a Sink for Waste CO2

Authors:
Benson, T. J., Lamar University
Madugula, A., Lamar University
Hovorka, S. D., University of Texas
Meckel, T., The University of Texas at Austin

Research and implementation on Carbon Capture and Storage (CCS) technologies have increased significantly in recent years and are showing that this technology can be viable in limiting atmospheric release of anthropogenic CO_2. Carbon-intensive industries along the U.S. Gulf of Mexico include two-thirds of the U.S. petroleum refining capacity and other large point source CO_2emissions sites. Matching deep subsurface geologic storage sites to the point sources is needed to develop large-scale implementation of CCS. The Gulf of Mexico is one of the most well-studied geologic basins in the world due to decades of petroleum exploration. Large volumes of prospective deep (>1 km) subsurface storage have been identified in State waters within 10 miles of the shoreline.

Through a grant by DOE-NETL, the Gulf of Mexico Partnership for Offshore Carbon Storage (GoMCarb) has been developed to assess offshore, sub-seafloor geologic CO_2storage in the Gulf of Mexico region. Early studies indicate that offshore storage is low-risk and likely offers near-term solutions to anthropogenic carbon emissions. Examination of seismic data shows that the stratigraphy of the area has excellent capacity to trap $CO_2$for permanent storage. GoMCarb seeks to identify suitable storage sites as well as required infrastructure (i.e. pipelines, and seafloor or platforms injection facilities). Risk assessment and mitigation tools are also being developed as part of the program. This talk will provide an overview of the GoMCarb project and will give an update of the latest findings of the research.

Retrofitting an Australian Brown Coal Power Station with Post-Combustion Capture

Authors:
Gulen, J., Bechtel
Elliott, B., Bechtel
Benz, G., Consultant
Qader, A., CO2CRC
Gibbins, J., UK CCS Research Centre
This presentation includes work performed by Bechtel in executing a study for CO2CRC Limited, Australia, on Post Combustion Capture (PCC) facilities at a reference brown coal power plant in the Latrobe Valley.

The study includes the conceptualization of a modular PCC Plant and associated Order-of-Magnitude Capital (Level 5 AACE).

The base case scope of each PCC module is an open-art absorption/stripping/compression treatment scheme using 40 wt% monoethanolamine (MEA) as absorbent. Fairly uniquely, the use of an open access design means the PCC scheme can be described in detail.

Performance and cost data are detailed for a single PCC module and, importantly, a concept to retrofit all of the units on site is presented. There are four boiler units at the reference plant, each including two exhaust gas flues with one PCC module per flue. Complete retrofits of multi-unit sites will obviously be required in the future, but to date both studies and PCC plants built have concentrated on single units.

Any PCC retrofit study will be specific for the reference site only and will include unique features. In this case the approach used had to allow for features including the absence of FGD units, the particular steam cycle components and layout, and connections to the unit stacks.

Retrofits, particularly for solid fuel plants, must also be supported by thorough pilot-scale testing and an appropriate project development process. This will also be discussed.

While the results of the study apply only to the reference plant, the general approach is of interest for wider application. In particular, the use of open access as an effective way to reduce overall costs of ownership for PCC retrofits will be examined.

Carbonation Processing and Strength Evolution of Portlandite-Based Cementing Systems

Authors:
Sant, G., University of California, Los Angeles
Mehdipour, I., Laboratory for the Chemistry of Construction Materials (LC2), UCLA
La Plante, E., University of California, Los Angeles
Falzone, G., UCLA
Pilon, L., University of California, Los Angeles
Neithalath, N., Arizona State University
Innovative binders based on portlandite ($Ca(OH)_2$) compositions have the potential to take up carbon dioxide (CO_2), e.g., from flue gas streams, and form calcium carbonate compounds which offer cementitious attributes. But, carbonation processing of such materials requires elaboration of how complex pore networks, and water saturation therein affect CO_2 transport and uptake. Therefore, this paper elucidates how the creation of air-filled porosity resulting from processing conditions regulates CO_2 mineralization and carbonation strengthening in portlandite-enriched cementing formulations. Toward this end, the relation between processing and carbonation kinetics are elucidated and linked to strength development and microstructure evolution resulting from *in situ*CO_2 mineralization and subsequent pozzolanic reaction. In general, the extent of air-filled porosity scales with carbonation kinetics so long as a critical liquid water saturation ($S_{w,c}$) is exceeded; below which carbonation is sharply hindered. It is also shown that the formation of hydration products strongly hinders carbonation by: (i) decreasing the CO_2 transport properties, and, (ii) reducing the exposed surface area of carbonatable reactants through surface coverage. Finally, the extent of carbonation is correlated with strengthening vis-Ã -vis uncarbonated systems that utilize traditional silicate cement hydration. An unprecedented level of strengthening â€" of â‰¥ 3.5x â€" is achieved for dry-cast (carbonated) portlandite mortars. On the basis of these concepts, the processing-carbonation kinetics relations in portlandite-enriched cementing systems enable exploiting prehydration, *in situ* CO_2 mineralization, and pozzolanic reaction, featuring enhanced early- and later-age strength development. These outcomes offer new insights that are needed to create novel pathways for utilizing *in situ* CO_2mineralization as a means for designing carbonate-based, CO_2 sorbing cementing agents for construction applications.

Low-Temperature Synthesis of Portlandite for Carbon Dioxide Mineralization

Authors:
La Plante, E. C., University of California, Los Angeles
Sant, G., University of California, Los Angeles
Castaño, S. V., University of California, Los Angeles
Neithalath, N., Arizona State University
Pilon, L., University of California, Los Angeles
The incorporation of CO_2 within stable metal carbonates such as calcium and magnesium carbonates is one strategy for the mitigation of atmospheric CO_2concentrations. The precipitation of $CaCO_3$ (e.g., calcite) from a Ca-rich solution requires the consumption of alkalinity and is favored by high Ca concentrations and high pH. The alkaline Ca salt, portlandite ($Ca(OH)_2$) is a potent CO_2 sorbent that can uptake 0.59 g CO_2 per g $Ca(OH)_2$ through the net reaction: $Ca(OH)_2$ + CO_2 â†' $CaCO_3$ + H_2O. However, current means of portlandite production entails a high energy consumption and high CO_2 emissions from the calcination of limestone (primarily $CaCO_3$) at elevated temperatures of about 800 Â°C to form lime (CaO), which is subsequently reacted with water to produce $Ca(OH)_2$. In this study, we develop a low-temperature process for portlandite synthesis which does not involve calcination and the associated release of CO_2. The precipitation of portlandite at sub-boiling temperatures is induced through the steps of: Ca extraction, Ca concentration enhancement, and portlandite crystallization via temperature ramping. A Ca-rich solution obtained by dissolving crystalline slag in deionized water is filtered through a membrane which selects for the divalent Ca^{2+}ions to increase the solution Ca concentration, thus reaching portlandite saturation. Upon reaching equilibrium with respect to portlandite at 25 °C, the solution is heated to temperatures of up to about 70 °C. Because the solubility product of portlandite decreases with increasing temperature, the temperature increase causes the precipitation of the phase out of the solution. Solid portlandite yields corresponding to the theoretical value were obtained. Moreover, the total energy consumption and CO_2 footprint of the process were estimated and compared with those for the traditional production method, revealing that the presented alternative route for the synthesis of portlandite features a potentially smaller environmental impact.

Nano-Engineered Catalyst for the Utilization of CO2 in Dry Reforming to Produce Syngas

Authors:
Li, S., GTI
Jin, B., Missouri University of Science and Technology
Shang, Z., Missouri University of Science and Technology
Liang, X., Missouri University of Science and Technology
The objective of this study was to develop a novel catalytic reactor containing nano-engineered catalyst for the utilization of CO_2 (captured from coal-fired power plants) in dry reforming of methane (DRM) (CO_2 + CH_4 â†' 2 H_2 + 2 CO) to produce synthesis gas (syngas). The technology aims to reduce CO_2 emissions by developing beneficial uses for CO_2 from coal-fired power plants. It also offers an alternative to mitigate CO_2 emissions in areas where geologic storage may not be an optimal solution and/or utilization could significantly offset the costs of carbon capture and sequestration.

The nano-engineered Ni-based catalyst was prepared by atomic layer deposition (ALD). The Ni particles were as small as ~2-4 nm. The nano-engineered catalyst showed CH_4 conversion >95%, H_2/CO ratio in the range of 0.7-1.0, and CH_4reforming rate as high as 2,500 L/h/g_{Ni} at 850 Â°C and pressure of 15-25 psia. The Ni-based ALD catalyst also showed good stability in DRM reaction during a 200-hour continuous operation at 850°C. This is due to a strong bonding between the nanoparticles and substrates since the Ni nanoparticles were chemically bonded to the substrate during the ALD process. The high thermal stability maintains high dispersion of Ni nanoparticles, which can inhibit coke formation, because their step edges are small enough to limit carbon nucleation and growth.

Performance Verification of Carbon Utilization Technologies in the Nrg Cosia Carbon Xprize, Transforming CO2 into Valuable Products

Authors:
Chatterton, B., Southern Research
McCabe, K., Southern Research

The $20 million NRG COSIA Carbon XPRIZE is a global competition to develop breakthrough technologies that will convert COâ„ emissions from power plants and industrial facilities into valuable products like building materials, alternative fuels and other items that we use every day. Under contract to XPRIZE and following ISO compliant protocol, Southern Research provides independent/unbiased field performance evaluations of the technologies/teams that have entered the competition.

The NRG-Cosia Carbon XPRIZE competition consists of 3 rounds with down-select of the most promising technologies after each round. The competition is structured as a two-track prize, with one track focused on testing technologies at a coal power plant, and the other focused on testing technologies at a natural gas power plant. 47 teams submitted proposals to XPRIZE and entered Round 1 of the competition. From these entries, 27 were down-selected for Round 2 by an independent panel of experts that was appointed by XPRIZE. In Round 2, Southern conducted independent performance verification of the technologies at laboratory scale. Results of these assessments were used by the judging panel to further down select the field to 10 finalists for long term pilot scale demonstration â€“ 5 at a coal fired power plant (the Wyoming Integrated Test Center) and 5 at a gas generating facility (the Alberta Carbon Conversion Technology Centre).

Southernâ€™s activities supporting the competition finals will include:

- On site verification activities during process commissioning and startup of each system,
- On-site short term operational verification,
- Off-site long term performance validation,
- Final assessments of all data and comparison to XPRIZE Round 3 requirements,
- Review of submitted final data and reports from each team, and
- Generation of final Verification Reports.

The final verification reports for each team will be used by XPRIZE and the judging panel to complete the competition.

Carbon Capture: The Scope of Carbon Capture Technologies in Indian Steel Industry

Authors:
Sangode, R. G., MECON Limited

The steel industry is among the most energy-intensive industries of manufacturing sector emitting significant volumes of carbon dioxide (CO2). On average, 1.9 tonnes of CO2 are emitted for every tonne of steel produced [1] . According to the International Energy Agency, the iron and steel industry accounts for approximately 6.7% of total world CO2 emissions.[1] The industries sector accounted for 21.7 percent of the total GHG emissions or 412.55 million tons of CO2 eq in the year 2005. In the year 2007, iron & steel production resulted in 6.2 percent emissions (117.32 million tons).[2] India have committed 33-35 per cent cut in carbon emissions by 2030 in from 2005 level in the target submitted to the United Nations Framework Convention (UNFCCC) on Climate Change for a global climate pact.[3] The paper discusses various carbon capture storage (CCS) and carbon capture utilization (CCU) technologies are being developed and implemented in pilot plants / commercial plants in Europe and USA ,which may be implemented on commercial scale in Indian steel industries. The Indian steel sector, due to its scale of production and energy intensive nature, represents a sizeable opportunity to reduce CO_2 emissions. At 2018 energy and emission intensities, this production would entail 200 $MTCO_2$-eq as emissions

Keywords: iron and steel, production, energy intensity, CO2 emissions, CO2 mitigation, CCS & CCU

CCUS Applied to Industrial Sources: A Significant Near-term Deployment Opportunity

Authors:
Munson, R., Global CCS Institute

There is growing acknowledgement that the capture and storage of CO_2emissions from industrial sources is an attractive opportunity for emissions reduction that is often easier and cheaper than capture from power plants. Industrial sources account for around 25 percent of global CO_2 emissions, and those emissions are expected to grow by over 50 percent by 2050. The International Energy Agency (IEA) has projected that, in order to limit global temperature increase to less than 2ËšC, almost half of the CO_2 captured and stored between 2015 and 2050 would come from industrial sources. Much of the progress that has already been made on CCUS is associated with industrial sources. Of the 20+ large-scale CCS facilities currently in operation or under construction globally, 19 capture CO_2 from industrial sources. The majority of those include facilities where the CO_2 is already separated as part of the production process (e.g., natural gas processing and fertiliser production) and is relatively inexpensive to capture. These 19 industrial facilities have a CO_2 capture capacity of approximately 35 million tonnes per year. While industrial sources represent a significant percentage of global CO_2 emissions, there are challenges associated with widespread deployment of capture technologies. Much of the governmental and public attention on reducing CO_2 emissions has focused on the power sector, reducing the sense of urgency for development of cost-effective approaches in industrial sectors. In addition, the products of the industrial sector are often competing in a global market with tight profit margins, making manufacturers sensitive to changes in production costs. Finally, emissions are scattered over many industries with widely divergent characteristics. This presentation will provide a summary of current industrial CCUS activity, associated costs, and opportunities for future deployment given incentives such as the 45Q policy.

Characterization of a World Class Carbon Dioxide Storage Complex in Kemper County, Mississippi, USA

Authors:
Riestenberg, D., Advanced Resources International, Inc
Koperna, G. J. Jr., Advanced Resources International, Inc.
Pashin, J. C., Oklahoma State University
Sams Gray, K., Southern States Energy Board
Esposito, R., Southern Company

The southeastern United States contains the nationâ€™s largest CO_2 emission point-sources, and the regional geology hosts a suite of structures with considerable CO_2 storage potential. The *Project ECO$_2$S* storage site encompasses 12,000 hectares adjacent to Mississippi Power Companyâ€™s (MPC) Kemper County energy facility in Mississippi, U.S.A. The stratigraphy of the site consists of stacked storage reservoirs and baffles with thick overlying seals. A comprehensive research team was assembled from 17 universities, national laboratories, non-profit institutes, and private companies to assess the storage potential and petrophysical parameters at the site.

The project focuses on assessing a broad suite of reservoir and confining properties for the Paluxy Formation, the Washita-Fredericksburg interval, the Lower Tuscaloosa Group â€˜Massive Sandâ€™ and the â€˜Marineâ€™ Tuscaloosa from three characterization wells drilled in the summer of 2017. Project objectives focus on: 1) confirm storage reservoir volumetric properties and develop a dataset on flow parameters; 2) advanced core tests; 3) caprock studies; 4) descriptions of depositional facies, rock types, and mineralogies of storage reservoirs and caprocks; 5) rock mechanics modelling; 6) fluid-rock geochemical interaction modelling; and 7) an evaluation of existing 2D seismic data.

In conjunction with geotechnical assessment of study area, the successful development and execution of a large-scale CO_2 storage project necessitates a strong risk management plan. A workshop was conducted to identify and evaluate the principal risks. Identifying and evaluating every important source of risk is of key interest to future owners/operators of such a project. The results of these analyses will inform a reservoir simulation of commercial scale CO_2 storage and a project risk assessment.

Project ECO$_2$S is part of the CarbonSAFE Program and is financially supported by the USDOE-NETL and Mississippi Power Company. The project is managed by the Southern States Energy Board. Technical Support is provided by Southern Company Research and Development.

Moving Bed Chemical Looping Technology for Carbonaceous Fuel Combustion and Hydrogen Production: Process Development and Assessment

Authors:
Tong, A., The Ohio State University
Xu, D., The Ohio State University
Zhang, Y., The Ohio State University
Wang, D., The Ohio State University
He, P., The Ohio State University
Hsieh, T. L., The Ohio State University
Guo, M., The Ohio State University
Nadgouda, S., The Ohio State University
Velazquez-Vargas, L. G., Babcock & Wilcox Company
Flynn, T., Babcock & Wilcox Company
Fan, L. S., The Ohio State University
Chemical looping technologies are a promising alternative for the efficient conversion of carbonaceous fuels to electricity and/or high value chemicals with minimal carbon emissions. These processes utilize an oxygen carrier solid material to indirectly supply oxygen to the fuel source. Ohio State applies the chemical looping concept for power and H_2 production from coal and natural with in-situ CO_2 capture. Specifically, Ohio State has developed the coal direct chemical looping (CDCL) process for power production applications from coal and the syngas chemical looping process (SCL) process for high purity hydrogen production from gasified coal or natural gas with nearly 100% CO_2 in both processes. The counter-current moving bed reactor design the CDCL and SCL process ensure nearly full fuel conversion to CO_2 with minimal solid circulation and capability of producing high purity H_2. The reduced metal oxides from the reducer operation are regenerated via an oxidation reaction with air. Various active metal oxides or metal oxide composites can be utilized. A high pressure 250 kW_{th} SCL pilot plant was constructed and tested at the National Carbon Capture Center proving the capability of the moving bed process for hydrogen production. A 250 kW_{th} CDCL pilot plant was constructed at the Babcock & Wilcox Research Center showing nearly full coal conversion with nearly 100% CO_2 capture. The present presentation summarizes the oxygen carrier particle development for both systems, the operational results from the 250kW_{th} CDCL and SCL pilot plant operations. Key process design features, process performance analysis, and techno-economic assessments will be discussed. Results from recent demonstrations and process assessments show how the chemical looping concept has the potential to transform the way carbonaceous fuels are used today to produce power and high value chemical products.

2018 EOR Survey in China-Part 2

Authors:
Guo, H., Yan'an University
This paper provides detailed field scale CO2-EOR survey in China. Part 1 has been orally presented in SPE organized Improved Oil Recovery conference held in 2018. All CO2-EOR field tests as well as two commercial application in China has been reviewed and summarized in detail. Lessons learned and opportunities in CO2-EOR are also introduced.

Since the first CO_2 huff-n-puff pilot was carried out in Daqing in 1963, 44 CO2-EOR field tests have been conducted in China. 26 have been conducted by SINOPEC, while 16 have been conducted by CNPC. 2 pilots have been conducted by the only local oil producer in China. The largest commercial CO2 flooding is in Jiling Oilfield, CNPC and the annual injection CO2 volume is 0.5 million tons. CO2 is from CO2 gas reservoir nearby. CO2 flooding in Daqing Oilfield (CNPC) is on semi-commercial or industrial stage, indicating that CO2-EOR is feasible. A large semi-commercial CO2 flooding project with 9 injectors and 38 producers in under construction after two well injectivity tests finished in 2018 in a ultra-low permeability reservoir in Changqing Oilfield (CNPC). The incremental oil recovery is predicted 15% OOIP, while the current field water flooding recovery is 18.5% OOIP. CO2 steam flooding pilot has been tested in Xinjiang Oilfield (CNPC) in January 2019 and annual CO2 injected is 100 000 tons captured from a factory. As for SINOPEC, CO_2 flooding in Huadong, Shengli, Zhongyuan are on commercial application stage. CO2 channeling and corrosion are problems to be solved. CO2 cost from chemical plants is as low as 140 RMB/ton. Oil production from CO2 flooding accounts for less than 1% of total EOR oil in China.

It is the first time that all CO2 EOR tests in China are surveyed in detail. Interest in CO2-EOR is increasingly rapidly recently in China.

A New Evaluation Method of CO2 Storage Capacity in Coalbed Reservoirs

Authors:
Yao, S., PetroChina
WU, M., PetroChina
Carbon Capture Utilization and Storage (CCUS) is playing a significant role in dealing with the global climate change. In order to study CO_2 storage capacity in coalbed reservoirs, adsorption/desorption experiments of coal rocks from different regions and coal ranks are conducted. Experimental results show that laws of desorption and absorption for two different coal ranks are obviously different and CO_2/CH_4 adsorption ratio is different along with reservoir depth. This paper points out weaknesses of evaluation methods in existing models, shows clearly what problems existing in the previous methodologies and regards the proved ratio is a key factor when evaluation. Various factors are taken into account in the new evaluation method including proved ratio, replacement ratio of CO_2 to CH_4, recovery ratio, ash content and moisture, completion factor, efficient pores volume, gas saturation and so on. Therefore, it is the first time to clearly put forward storage depth limitation. In addition, new methods for evaluating theoretical, effective, practical and matched capacity are proposed. Also, reasonable depth of CO_2 storage is determined in China according to present production technology. Finally, this paper expresses evaluation results of CO_2storage capacity in Chinaâ €™s coalbed reservoirs which prove the applicability of new methodologies.

Assessment of Forest Degradation in Omo-Sashsa-Oluwa Forest Reserves, Southwestern Nigeria

Authors:
Eke, J., National Open University Nigeria
Samuel, O., National Open University
Forest degradation is a major environmental problem in many developing countries within the tropics. The Omo- Shasha- Oluwa forests is one of the most densely populated areas in Africa, and already shows high levels of human activity. The study intends to investigate changes in forest integrity due to anthropogenic activities by assessing flora biodiversity in study area and determining the trend of degradation using time series mapping. Stratified random sampling technique will be used to select sampling plots. 0.5km by 0.5 km grid will be overlaid on the land use land cover map and 5 representative plots each for the vegetated land use classes will be selected. Garmin 78 hand held GPS with an accuracy of 3 meters will be used to locate the center of the selected plots. The concentric plot assessment system will be used to select sub plots. Each modified plot is a cluster of four circular 17.95 m radius annular plots with one central 0.1 ha annular plot, three satellite 0.1 ha annular plots, four 7.32m subplots, and one 2.77 m radius micro plot. Each modified plot also contains three 17.95m long transects from the cluster center, with the first transect positioned at a random azimuth and the others at 60° and 120° from the first transect. Complete vegetation assessment will be carried within the sampled plots during field survey. Maximum likelihood algorithm will be used for the supervised classification after which validation and accuracy assessment will be conducted. Land change detection will be carried out using suitable GIS software. The data collected in situ from the field will be subjected to appropriate statistical analysis to determine the impact of degradation on the tropical forest. The result will provide invaluable information on the current landuse dynamics in the forest reserves of Southwestern Nigeria.

Carbon Utilization and Storage of Associated and Enriched Gas

Authors:
Konstantinos, K., University of houston
Salman, M., University of Houston
Sharma, P., University of Houston
In recent years, gas injection has become the preferred enhanced oil recovery (EOR) method onshore USA, with CO_2 being the most widelyâ€“used gas with the added benefit that a greenhouse gas (GHG) is sequestered. Much research has gone into improving CO_2 floods, in areas of conformance control. Improving conformance and mobility control will improve sweep efficiency which increases the oil recovery ratio. The use of CO_2 has some drawbacks: it can lead to significant corrosion issues; the supply may not always be commercially accessible, especially to fields located far from gas supply pipelines. Therefore, there is a strong driver to investigate alternative gas injectants.

Hydrocarbon gasses are more attractive for ease of handling, availability, and economic value. In many cases, associated gas (a GHG) that is produced at a field is vented or flared, when it could be utilized for EOR. To date, limited research has been conducted on possible enhancements to simple hydrocarbon gas injection, such as creating foam with surfactants and injecting gas in supercritical condition. To explore this potential, displacement tests were performed using ethane and compared with carbon dioxide. Aqueous stability, salinity scan, and static foam tests were performed to identify a surfactant formulation suitable for use with ethane. Foam quality and coreflood displacement tests in heterogeneous carbonate outcrop cores were conducted to compare the recovery efficiencies of three displacement processes. Our work shows that miscible ethane injection processes can result in significant recovery even under gravity-unstable conditions. The addition of foam further enhances overall recovery at laboratory scale. Results demonstrate that miscible ethane foam can be generated to address conformance issues and could lead to the wider use of associated gas for EOR while sequestering a GHG.

Co-Optimization of CO2 Storage and Oil Recovery Using Alcohol-Treated CO2 Injection

Authors:
Saira, S., University of New South Wales
Hussain, F., University of New South Wales
Ajoma, E., University of New South Wales
CO_2 sequestration into oil reservoir is an attractive method because of enhanced oil recovery associated with it. Combining both sequestration and oil recovery in a project add more economic value. This paper investigates effect of alcohol-treated CO_2 on co-optimization of CO_2 storage and oil recovery under immiscible and near miscible conditions.

Injection experiments were performed on a Bentheimer sandstone for two sets of fluid pairs (1) CO_2 and oil, and (2) alcohol-treated CO_2 and oil. A mixture of 35% Hexane and 65% Decane was used as oil. Alcohol-treated COâ,, prepared by mixing of 4% Ethanol in scCOâ,, at the experimental pressure and temperature. All experiments were performed at a constant temperature of 70 °C and two different pressures (1300 and 1700 psi). Literature studies show that 1300 psi represents immiscible condition and 1700 psi represents near-miscible condition for the selected oil and COâ,,. Oil recovery, differential pressure, and compositions are recorded during injection experiments. A co-optimization function for CO_2 storage and incremental oil production is calculated by use of measured data for each experiment.

Experimental observations reveal that CO_2 storage efficiency and ultimate oil recovery is increased 10-12% and 12-15% respectively with the addition of 4% ethanol in CO_2. However, results suggest that degree of improvement in CO_2storage and oil recovery with alcohol-treated CO_2 is more dominant during immiscible condition. Although it also shows improvement at near miscible condition, its efficiency is greatly reduced.

Conversion of Captured CO2 from Post-Combustion Coal Fired Power Plants to Useful Chemicals

Authors:
Balaja, O., Prairie View A & M University
Dada, E. A., ChemProcess Technologies (CPT), LLC

Carbon dioxide (CO_2), is the greatest contributor to the greenhouse effect without any doubt is one of the leading detrimental gaseous compounds our planet faces. Global warming resulting from the direct effect of the release of carbon dioxide into the environment in the recent years has been a topic of interest among scholars and nations. Carbon dioxide has also been considered as the main cause of global warming constituting approximately 55% of the world total emissions (Dermirbas 2005). The idea embraced in this paper focuses majorly on the economically and industrially viable conversion processes of carbon dioxide into useful chemicals, some of which are: urea, ethanol and electro fuels. An insight into the postâ €“combustion carbon capture technologies that are used in coal fired power plants and how the captured carbon is converted into various useful chemicals is examined. The main technologies adopted in pre-combustion, post combustion and how the captured carbon dioxide serves as feedstock for making useful chemicals are discussed in terms of performance, effectiveness, sustainability and compatibility to current technological trends. An attempt would be made to retrofit an optimized carbon dioxide conversion unit to a preâ€“existing coal fired power plant. A cost effective and efficient industrially viable carbon dioxide conversion processes might just be the route to the substantial utilization of carbon dioxide and drastically reducing the negative environmental impact of carbon dioxide on the environment. An outlook is given of other less economically viable developments in the carbon capture and conversion systems, primarily still in the laboratory scale. The knowledge gap in these areas centers on the operational costs of carbon dioxide conversion while other processes struggle to balance the reaction catalyst needed to attain descent carbon dioxide conversion rate. An extensive research particularly in the pilot scale area is recommended.

Environmentally Safe Offshore Platforms

Authors:
Sherbakova, K., Russian State Geological Prospecting University

During E&P production, a massive value of associated gas is being burnt and wasted. Such flare burning of associated gas - non-organic components create outburst of carbon dioxide (CO_2) and leads to environmental pollution and Green House Effect. Moreover, upstream traditionally powered by fuel oil generators which adds negative outburst of carbon dioxide (CO_2) into the environment and leaves negative effect on marine life and fauna. Minimization of outburst of associated gas during E&P though implementation a hybrid renewable energy system based on two types of energy sources combined: separation of CO_2 from associated gas by striping it out from gas and natural waveâ€™s movement. Creation of hybrid renewable energy system provides opportunity eliminate outburst of CO_2 from associated gas and make profit from wast burned flare though generation of alternative electrical pure energy from nature without negative outburst of carbon dioxide. Both components of hybrid system work together and complement each other. Through chemistry process, combining pure electrical power from waves and CO_2 can provide new business opportunity for recycling and utilization of CO_2 from flare outburst.

Create hybrid model of renewable energy generator during E&P by combining sources from Ocean waves and flare filtration of negative outburst of carbon dioxide (CO_2) by implementation of "process station" on the Jack-up rig site. The first component of the hybrid power source of renewable energy system shall be considered constant wave energy implemented into system of several non-sinking buoys on the water surface. Buoys are continuously registering and counting even slight movement of waves and produce pure electrical power. Accumulated electrical energy from wave generator by itself may solve only partial problem of CO_2 outburst by contributing powering energy utilized for rig and minimizing usage of fuel oil generators. This contribution is minimal with low efficiency rates.

Evaluation of Ccus Pilots in China

Authors:
Guo, H., Yan'an University
Lv, L., Yan'an University
Although many CO2 Capture, Utilization, Storage (CCUS) has been studied in laboratory, few has been tested in actual field in China. Evaluation of previous typical CCUS pilots are presented. CO2-EOR is the most attractive CCUS technology in China because it improves the benefits and reduces the cost. Up to present, 44 pilots have been conducted in China. 26 CO2-EOR pilots have been conducted by SINOPEC, while 18 by CNPC. 6 typical pilots conducted in Jilin, Daqing, Shengli, Huadong, Zhongyuan and Jinbian are both technically and economically evaluated. CO2-EOR pilots are both technically and economically successful and CO2-EOR in Jilin Oilfield is most mature in China since CO2 flooding is used commercially. Up to April 2019, there are five demonstration CCS-EOR projects in Jilin, covering 65 injectors and 318 producers. Annual CO2 injection is 1.42 million ton and incremental oil is 131 thousand tons. Incremental oil recovery factor is 20% OOIP. 2 large scale semi-commercial CO2 flooding tests (Bei 14 and Shu 101) are active in Daqing low permeability blocks and CO2 flooding oil accounts for 10%. Gao89-1 in Shengli is the largest scale test in Shengli. 24 wells are involved and the technical result is not as good as expected. CO2-EOR in Zhongyuan is very attractive since captured CO2 from chemical plants is injected. Produced CO2 is reinjected into reservoir. 1million CO2 is captured. The economical performance is good. CO2-EOR in Jinbian is not as good as expected, due to insufficient CO2 injected and no CO2 capture device at wellhead. The only CO2 storage in saline layer project carried out by Shenghua was stopped after injection 300000 tons CO2 for three years. The only CO2 enhanced coalbed methane (CO2-ECBM) pilot finished in 2004 got technical success but no economic success. 99% injected CO2 is storage in coalbed.

Flourinated Membranes for CO2 concentration and Carbon Capture

Authors:
Murnen, H., Compact Membrane Systems
Shangguan, N., Compact Membrane Systems
Majumdar, S., Compact Membrane Systems

The capture of CO_2 from power plants and industrial exhaust gases could significantly reduce CO_2 emissions. Flue gas streams are at atmospheric pressure and typically contain 4â€"20% CO_2 in mostly N_2. The objective of the separation process is to produce >99% pure liquid CO_2, at a pressure of 100â€"150 bars at a cost of $40 â€" 60/ton of CO_2 captured. Membranes have been suggested as a cost-effective method to concentrate teh CO2 found in flue gas streams to allow condensation to liquid CO_2. Because of the prohibitive cost of compressing the flue gas feed stream to generate a driving force across the membrane, vacuum operation on the permeate is more cost-effective. However, the pressure difference across the membrane is no more than 1 bar, so very high permeance membranes must be used. Todayâ€™s best membranes made from rubbery polar polymers have a CO_2 permeance of about 2000 GPU and a mixture CO_2/N_2selectivity of 20â€"30 at operating conditions. Membranes with higher permeances would help reduce cost. For example, a membrane system for a typical 500 MW_e coal power plant is likely to have ~1 million m^2 of membranes, which would be a significant fraction of the plant cost. A doubling of membrane CO_2 permeance without selectivity loss would reduce the required membrane area by half, significantly reducing cost and increasing the competitiveness of the membrane process.

Compact Membrane Systems (CMS) has developed high permeance custom amorphous fluoropolymer (CAF) membranes for selective concentration of $CO_2$from power plant flue gas and other industrial gases. The initial CAF results for concentrating CO_2 are very encouraging with very high permeances >2500 GPU. The CAF membranes also have high chemical resistance and high anti-fouling capabilities due to their highly fluorinated nature. This presentation will discuss our promising membrane results plus downstream engineering evaluation.

Investigations of Carbon Sequestration Mechanisms in Partially Depleted Oil Reservoirs

Authors:
Kutsienyo, E. J., New Mexico Institute of Mining and Technology
Ampomah, W., New Mexico Institute of Mining and Technology
Sun, Q., New Mexico Institute of Mining and Technology
Balch, R., New Mexico Institute of Mining and Technology
Cather, M., New Mexico Institute of Mining and Technology

The objective of this work is to evaluate various CO_2 trapping mechanisms and long-term carbon storage potential involved in CO_2 enhanced oil recovery (EOR) projects in a partial depleted sandstone oil reservoir. More importantly, we seek to confirm changes in petrophysical properties induced by CO_2 injection that have been observed from laboratory measurement. A field-scale numerical simulation model is established utilizing extensive field data to study the Upper Pennsylvanian Morrow B sandstone. The validity of the reservoir model is confirmed via a rigorous history matching study using field injection and production data. The history-matched model is employed to assess the multiple trapping mechanisms including residual trapping, structural-stratigraphic trapping, solubility trapping and mineral trapping. Numerical experiments are conducted considering the intra-aqueous and mineral dissolution/precipitation reactions, which may promote pore structure alterations. Moreover, we take advantage of Henryâ€™s law to calculate the CO_2 dissolution as function of pressure, temperature and salinity. The reservoir is monitored for 1000 years after all the injectors and producers are shut-in to investigate the long-term storage of CO_2injection in the field. Such a robust reservoir model is compentent to evaluate long-term CO_2 sequestration potential and study the impact of the CO_2 injection on petrophysical properties of the Morrow-B formation, such as pore fluid composition, mineralogy, porosity and permeability. The results from this work provide significant insights in terms of the optimum storage of CO_2 in aqueous-gaseous-mineral phases, effect of salinity on storage, and the amounts of dissolution/precipitation of the principal accessory minerals evolving throughout the project. Information gained from this study offers valuable insight regarding physiochemical storage induced by the CO_2 injection activities and may serve as a benchmark case for future CO_2 -EOR projects when reactive transportations are considered.

Laboratory Experiment of CO2 Huff and Puff in Tight Reservoir and Determination Method of Development Parameters for CO2 Huff and Puff Development

Authors:
LI, J. Sr., Exploration?Development Research Institute,Jilin Oilfield Company,PetroChina
Wu, W. Sr., Exploration?Development Research Institute,Jilin Oilfield Company,PetroChina
Wang, H. Sr., Exploration?Development Research Institute,Jilin Oilfield Company,PetroChina
Zhao, X. Sr., China university of petroleum(beijing)
CO2 huff and puff is one of the effective methods to improve reservoir recovery. For tight reservoirs, the research on the development mode and development parameters of CO2 huff and puff is insufficient. Through the laboratory experiments of CO2 huff and puff, the difference between the mechanism of CO2 flooding and that of CO2 huff and puff was analyzed. The relationship between the amount of CO2 injection and the solubility of CO2 in formation crude oil was studied. The influence law of the sweep volume of CO2 by the stuffing time was clarified. Finally, the method of determining the development parameters of CO2 huff and puff was established. The results show that the main mechanism of CO2 flooding to improve oil displacement efficiency is to reduce interfacial tension and displacement resistance, while the mechanism of huff and puff is to maximize the solubility of CO2 in crude oil, so it is not necessary to form miscibility in CO2 huff and puff. When CO2 injection rate is lower than CO2 dissolved in crude oil, better oil displacement effect can be obtained. If the injection rate is too high, the development effect will become worse. The development effect of CO2 huff and puff in closed reservoir is better. If the reservoir is not closed, the development mode of small injection volume, multiple rounds and supplementary supply can be adopted

Monitoring Utilizing Multiple Data Sets for the Carbon, Capture, Storage, and Utilization Project at Farnsworth Unit, TX

Authors:
Czoski, P., New Mexico Tech
Balch, R., New Mexico Institute of Mining and Technology
El-Kaseeh, G., New Mexico Tech

The Southwest Partnership on Carbon Sequestration (SWP) is a CO_2 carbon capture utilization and storage (CCUS) project sponsored by the U.S. Department of Energy, with a goal of permanently storing and monitoring 1,000,000 tonnes of CO_2. The SWP project is located in a mature field undergoing enhanced oil recovery (EOR) at Farnsworth Unit (FWU), Texas, USA. Utilized CO_2 is anthropogenic, sourced from a fertilizer and an ethanol plant. The field has 15 CO_2 injectors and has injected 1.5 million tonnes of dense phase CO_2, with over 900,000 tonnes stored, since 2013. Major project goals consist of generating best practices for CCUS and developing ways of characterization, optimization of production, and monitoring. To help ensure that the injected CO_2 is being stored in the reservoir, monitoring is required in multiple forms.

The ongoing monitoring program at FWU in the Morrow B reservoir includes geophysical methods such as time lapse vertical seismic profiling (VSP) and microseismic monitoring. The baseline VSP and three subsequent monitor surveys make plume tracking possible in one four-spot injection pattern. A detailed workflow integrating geologic interpretation, VSP time lapse studies, tracer tests, and subsequent recovery at producing wells aid in the understanding of flow controls in the reservoir. Microseismic activity has been identified on the original borehole array and more recorded events are expected on the new, low noise borehole array. As the reservoir is monitored, any additional information will be added to our geologic and simulation models in order to better assess the CO_2storage and production optimization.

Funding for this project is provided by the U.S. Department of Energy under Award No. DE-FC26-05NT42591. Additional support has been provided by site operator Perdure Petroleum LLC.

Optimization of Well Placement and Operating Condition for Various Well Types in CO2 Sequestration in Pohang Basin, Korea

Authors:
Kim, M., Inha University
Jun, C., Inha University
Shin, H., Inha University

Since CO_2 emission of South Korea has continuously increased, the government started a pilot project for geological storage into an aquifer in Pohang basin in 2013. Pohang basin has geological characteristics of multi-layered reservoir and pre-existing faults which restrict CO_2 injection to a certain amount. To store a large amount of CO_2 and keep the storage security, it is needed to define the optimum injection well type, placement and the operating conditions.

This study applied various well types to improve the capacity by changing well trajectories and controlling the pressure build-up, with monitoring the injection and fault pressure. Single vertical injection well, two vertical injection wells, single horizontal injection well, and two vertical wells for pumping and injection have been employed and designed in the numerical simulation using a geological model of Pohang basin. To determine the optimum well location and operating conditions of each well type, the optimization process was conducted with a commercial optimizer, Designed Exploration and Controlled Evolution (DECE) of Computer Modelling Group (CMG).

The results showed two vertical wells for pumping and injection at the optimum location and fluid rate kept the security within the pressure boundary and allowed almost 8 times larger volume of CO_2 to be stored compared to base case. When using a single well, vertical injection well injected 126 kton of CO_2 that was the largest, and horizontal well prevented the upward-migration of CO_2 for additional storage security. Through this research, it is expected to forecast the effect of various well types in the CO_2 sequestration in Pohang and propose the optimum well type and conditions for the largest storage capacity.

Performance of Rice Straw Ash As an Alternative, Low Cost Oxygen Carrier in Chemical Looping Combustion Technology

Authors:
Bhui, B., IIT Guwahati
Vairakannu, P., IIT Guwahati
Direct coal fuelled chemical looping combustion (CLC) is one of the efficient, advanced carbon capture techniques to generate CO_2 enriched flue gas for sequestration. This technology separates oxygen by the chemical reaction between metal particles and air and the resultant metal oxides can be used for fuel oxidation. Biomass ash contains a significant quantity of Fe, Ca etc. and this would be a promising source of metal oxides. In this context, rice straw based ash content (RSA) is used as the oxygen carrier for the CLC of low ash Indian coal (LAC) in a fixed bed reactor. Further, thermo-gravimetric analyses was carried out using a mixture of RSA and LAC under CO_2 and N_2 atmosphere from 25°C to 1000°C. The instantaneous reactivity of 3×10^{-4} s^{-1} and 6×10^{-4} s^{-1} was estimated in two distinct regions at the temperature intervals of 400-600°C and 800-1000°C, respectively, under N_2 atmosphere. This confirms the reactivity of metal oxide particles in the RSA with the coal. Also, a higher instantaneous reactivity of 1Ã—10^{-3} s^{-1} was found under CO_2 atmosphere between 800 to 1000Â°C. It is found that the RSA exhibited 3% higher mass loss, compared to the reactivity of pure Fe_2O_3 with coal char. Hence, the performance of RSA is found as equivalent to that of Fe_2O_3 oxygen carriers. X-ray diffraction analysis of RSA was conducted. It was found that Ca-ferrites ($CaFe_2O_4$ and $Ca_2Fe_2O_5$) and SiO_2 were present in the RSA. These components could be responsible for the CLC reactivity of coal as per the following reactions ($Ca_2Fe_2O_5$â†'2Fe+2CaO+1.5O_2, 5$CaFe_2O_4$â†'0.5$Ca_2Fe_2O_5$+$Ca_4Fe_9O_{17}$+0.25O_2) (Ismail et al. 2016). Hence, the results of the present study shows that RSA ash could be a potential oxygen carrier. However, further modifications are required to improve its durability such that it can be efficiently used over multiple cycles for commercial purpose.

Potential Application of CCS Technology for Reducing the Carbon Footprint Created during Deepsea Mining Operations

Authors:
Wu, S., SWu Houston LLC

Since some of the successful pilot tests carried out in the late 1970s, deepsea mining of high-value minerals, such as cobalt and rare-earth elements has been receiving renewed interest. The water depth range in which to operate a deepsea mining project is beyond 4,000m up to 6,000m. Because of this, deepsea mining operation will require a significant of power, the generation of which would create a significant carbon foot print even using IMO rule compliant marine fuel engine-generators. One potential technology that could be utilized to reduce the footprint is to capture and inject the CO2 from the exhaust stream of the mining vessel power plant deep down near the seabed where the density of the CO2 is significantly higher than that of the seawater. The discharged CO2 â€œplumeâ€ would sink down to form â€œCO2 Lakesâ€ in the troughs on the seabed. The key systems to be added onto the design of a deepsea mining vessel for the carbon capture are identified, but the main focuses of the paper will be on a riser configuration developed for the combined mining production and for the CO2 injection on a prospective deepsea mining project.

Rapid Evaluation Model of Enhanced Oil Recovery of CO2 Flooding Based on Modification Fractional Flow Theory

Authors:
Yu, Q. Sr., No.2 Oil Production Plant?PetroChina Changqing
Xi, H. Sr., No? 2 Oil Production Plant?PetroChina Changqing
Zhao, X. Sr., China university of petroleum(beijing)
Before a wide range of CO2 enhanced oil recovery (CO2 EOR) technology is implemented for an oilfield, the enhanced oil recovery potential and economy of every reservoir should be evaluated to determine which EOR method is most appropriate. Hundreds of reservoirs in an oilfield exist, and much work should be done. Thus, a rapid and reliable EOR evaluation model is necessary. Combining seepage theory with reservoir simulation, simulation models of CO2 flooding based on fraction theory are proposed in this paper. In the CO2 flooding model, the new minimum miscible pressure evaluation model is introduced to simulate miscible flooding and immiscible flooding, and the relative permeability is modified by the minimum miscible factor. The reduction factor of mobility ratio is introduced to modify the mobility ratio. Interfacial tension calculation formula is introduced to simulate the CO2 displacement mechanism. The simulation model is applied in a low permeability reservoir, and the simulation results show that this model presents certain reliability.

Reaction Experiments of CO2-Rock-Water-Crude Oil and Its Effect on CO2 EOR

Authors:
Yu, Q. Sr., No.2 Oil Production Plant?PetroChina Changqing
Zhao, X. Sr., China university of petroleum(beijing)
Xi, H. Sr., No? 2 Oil Production Plant?PetroChina Changqing
During CO2 flooding in reservoir, CO2 dissolves in water and forms to carbonic acid, which will react with formation water, rock and crude oil as follows: carbonic acid reacts with calcium and magnesium ions in formation water to produce inorganic salt precipitation, reacts with colloid and asphaltene to produce asphaltene precipitation, and reacts with rock to produce dissolution. The above reaction will affect the development effect of CO2 flooding. Through laboratory experiments, the mechanism and main controlling factors of inorganic salt precipitation formed by CO2 and formation water are analyzed, and the quantitative characterization model of inorganic salt precipitation is established considering temperature, pressure and ion content of formation water; the interaction mechanism between CO2 and crude oil is clarified, and the mathematical characterization model of asphaltene precipitation is established considering temperature, pressure and asphaltene content. Based on the mechanism of rock action, a mathematical model for quantitatively describing the amount of dissolution is established. Finally, the influence of inorganic salt precipitation, asphaltene precipitation and dissolution on enhanced recovery of CO2 flooding is analyzed by using reservoir numerical simulation method and taking typical reservoirs as an example. The results show that the distribution of asphaltene mainly concentrates near gas injection wells, but the total asphaltene has little effect on oil recovery; the precipitation of inorganic salt gradually increases from gas injection wells to oil production wells, and the precipitation of inorganic salt near gas injection wells is smaller, the dissolution effect near gas injection wells is stronger, and the effect of dissolution on oil recovery is greater.

Refiner's Industrial Scale Facility Uses CO2 in Stack Gas to Enhance Crop Production

Authors:
Kolodji, B., Black Swan, LLC
California's San Joaquin Valley is one of the nation's largest agriculture producing regions. As such at least one small oil refinery is uniquely located here and is surrounded by hundreds of acres of almond orchards and grape vineyards. This paper presents the science, patent pending gas processing technology, and other data supporting the use of this refinery's clean CO2 rich flue gas in neat, conditioned form to enable "crop carbon enrichment". With this technology, the science of crop carbon enrichment practiced worldwide in greenhouses and promoted for decades by the US Department of Agriculture (USDA) and The Brookhaven National Laboratory (BNL), is now commercially viable in an open air orchard (meaning without a greenhouse), making the conditioned flue gas a valuable commodity. The primary benefits are an increase in crop yield of up to 100% and an increase in efficiency of crop water utilization up to 30%, together resulting in nearly doubling the value of an orchard. Secondary benefits include a reduced carbon footprint for the refinery through the biosequestration of CO2 that is captured in the harvest yield and increased biomass of the plant root, trunk and leaves. Importantly, the use of crop carbon enrichment in this case shows significant value toward achieving California's goals for water conservation in the agricultural sector and the overall reduction in greenhouse gas emissions. The project is funded by a grant from the California Air Resources Board's (CARB's) California Climate Investment (CCI) Program as sponsored and awarded by the California Department of Food and Agriculture's (CDFA's) Office of Environmental Farming and Innovation (OEFI), State Water Efficiency and Enhancement Program (SWEEP).

Synthesis Metal-Doped Mesoporous Resin for Carbon Dioxide Capturing and Electrochemical Conversion

Authors:
Nabavinia, M., Rowan University

While global warming is a pressing issue in the news today, there still isnâ€™t enough being done to reverse the harmful effects of greenhouse gases like carbon dioxide. Lack of economically benefits for the removal of carbon dioxide is due to the absence of a promising catalyst. In this research, a single stage of the solvothermal method was used to synthesize mesoporous resin (MR) without any sacrificing templates. Resorcinol and formaldehyde used as a cost-effective precursor. The ethanolic solution of polyethyleneimine (PEI) was used to provide amide groups and mesoporous structure in resorcinol-formaldehyde template-free resin. Also, Cobalt (II) acetylacetonate and Ni (II) acetylacetonate were added to extend our studies in adsorption to the catalytic electrochemical conversion of CO2. BET surface analysis confirmed increasing of surface area by adding the ethanolic solution of PEI. This amid group been known to be excellent at capturing carbon dioxide, furthermore increasing surface area. MRâ€™s surface area is 325.75 m /g. The resulting material shows promising CO2 capture performance, reaching equilibrium adsorption of 1.1 mmol CO2 /g of material at 0 centigrade and a p of 1 bar. Also, metal ions not only increased surface area but also enhance carbon dioxide adsorption. The results showed cobalt one had the highest surface area and thermal stability (Co-MR (cobalt 3% w/w) BET surface area 675.62 m2/g, TG =450 centigrade).

Consequently, adsorption of 2.1 mmol CO /g of material at 0 centigrade a P of 1 bar exhibited. However, Ni-MR has highest electrochemical conversion. This mesoporous polymer is promising for CO capture and conversion due to their chemical inertness, low cost, high surface area, tunable pore structure, and ease of functionalization by the metal ion.

The Effect of Asphaltene Structure on Wettability Alteration during Enhanced Oil Recovery Using Carbonated Water

Authors:
Alqam, M., KFUPM
Alawani, N., Saudi aramco

The Effect of Asphaltene Structure on Wettability Alteration during Enhanced Oil Recovery Using Carbonated Water

The presence of the polar compounds in the crude oil play a major role in dictating the type of wettability. The objective of this study is to find out the effect of the structure of asphaltene and other polar compounds on wettability alteration during enhanced oil recovery using carbonated water. The presence of heavy components in the crude oil, for instance, asphaltene and resins play a major role in the wettability reversal.

Effect of pressure was investigated on crude oil /carbonated water/ pure Dolomite system using carbonated water (200 cm^3 liquid CO_2 mixed with 800 cm^3 of brine at the initial pressure of 2000 psi. On pure Dolomite solid surface was saturated with crude oil containing 11.6 wt% of asphaltene. Decreasing Dolomite content in mixed Dolomite/Calcite systems caused shift in contact angle from oil negative intermediate wet to weakly water wet regardless of saturating fluid phase.

The characterization of asphaltene using Fourier Transform Ion Cyclotron Resonance Mass Spectrometry (FT-ICR MS) has clearly indicated that asphaltene sample consist of condensed hydrocarbon and sulfur aromatic molecules with one or two sulfur atoms per molecule. The weight average number of fused aromatic rings for hydrocarbon molecules ranges from 6 to 7 and for sulfur containing molecules having an average of 8 aromatic rings,

Using FT-ICR MS, it was found that asphaltene contains almost similar number of poly-aromatic rings with different average alkyl chain length. However, the total S and N content was different in each oil sample. The more total S and N content leads to more polarity and more oil wet behavior. It seems that the difference wettability alteration is mainly rooted to the difference in polarity of the oil.

Understanding the Subsurface Structural Formation for Potential CO2 Storage and Capture in Geologic Formation

Authors:
Nwoke, U., Elvianna Project

Reduction of CO_2 to a minimal level to a time relative to the pre-industrial era required a huge volume of CO_2 sequestered. The question remains where could we potentially sequester these large volumes of CO_2? Generated by coal â€"fired power plants, Fossil Fuel, Cement Production factory, iron and steel and pulp and paper factory (Global CCS Fact Sheet, 2018). Studies as shown that three main types of geological storage options are generally considered for CCS purposes: deep saline formations; depleted oil and gas fields, and unmineable coal seams. Recently, additional potential opportunities have been shown to exist in gas shales, salt caverns and basalt formations, (Godec et al, 2013). The purpose of the research is to understand the structural, geological, thermodynamic, hydrodynamic, techno-economic, social potentials of various formation for the storage and capture of CO_2 in the geologic formation. With over 60% of worldwide emissions coming from point sources that are potentially amenable to CO_2 capture, the prospects for CCS to significantly reduce CO_2 emissions are great (IPCC 2005). Technical and economic assessments suggest that over the coming century, CCS may contribute up to 20% of CO_2-emission reductions, equivalent to reductions expected from efficiency improvements and large-scale deployment of renewable energy resources (IPCC 2005). At the pressures and temperatures expected for sequestration reservoirs, the volume required to sequester CO_2 as a supercritical fluid is about 10 million cubic meters (Mm3) per year. Sequestering the CO_2 emissions from a power plant with a 50-year lifetime would require a volume of about 500 Mm3(Sally M. Benson , 2008). It is therefore imperative that, suitable formations for storage potentials should be deeper than 800 m, have a thick and extensive seal, have sufficient porosity for large volumes, and be sufficiently permeable to permit injection at high flow rates without requiring overly high pressure.

3D-Printed Monoliths for CO2 Capture and Utilization Applications

Authors:
Rezaei, F., Missouri University of Science and Technology
Structured monoliths improve adsorption and reaction kinetics and reduce column pressure drops by increasing fluid contact area in comparison to traditional beads or pellets; however, targeted adsorbent/catalyst loading methods, such as *in-situ*growth on extruded monoliths, are costly and inefficient. 3D-printed monoliths can address these flaws by directly incorporating active materials into the structure matrix. Alternatively, 3D-printed monoliths can be doped with an adsorbent or metal catalyst precursor; leading to high internal loading. Herein, two novel routes for 3D-printed monoliths are *Direct Adsorbent Printing* and *In-situ Adsorbent Growth*. The former achieves high adsorption capacities with good mechanical stability while the later can achieve high loadings by incorporating a precursor directly within monolith matrix. Using these routes, we have successfully demonstrated the application of 3D printing technique for formulating a variety of materials including zeolites, aminosilicates, MOFs, and polymerâ €"zeolite composite monoliths as suitable adsorbent contactors to remove CO_2 from flue gas or enclosed environments[1-4]. Additionally, we have successfully demonstrated the development of 3D-printed ZSM-5 monoliths with various metal dopants for CO_2 utilization reactions, in particular, methanol to hydrocarbon (MTH) reaction[5].

The 3D printing technique offers an alternative cost-effective and facile approach to fabricate structured adsorbents and catalysts with tunable structural, chemical and mechanical properties for use in CO_2 separation and conversion processes. This work highlights the importance of adsorbent/catalyst materials formulations into practical contactors such as monoliths, as the scalabale technology platform, that could facilitate rapid deployment of CO_2 capture and conversion processes on commercial scales.

References:

1. S. Lawson et al. *ACS Appl. Mater. Interfaces 10*, 19076 (2018)
2. H. Thakkar et al. *ACS. Mater. Interfaces* 8, 27753 (2016)
3. H. Thakkar et al. *ACS. Mater. Interfaces* 9, 7489 (2017)
4. H. Thakkar et al. *ACS. Mater. Interfaces* 9, 35908 (2017)
5. F. Magzoub, et al. *Appl. Catal. B: Environ*. 245, 486 (2019)

A Review of the Mechanisms of Miscible and Immiscible Carbon Dioxide Injection in Shale Reservoirs for Storage and Increased Oil Recovery Applications

Authors:
Fakher, S., Missouri University of Science and Technology
Imqam, A., Missouri University of Science and Technology
Unconventional shale reservoirs are now considered a huge source of energy for the hydrocarbon industry. One of the currently proposed methods of increasing oil recovery from these reservoirs is carbon dioxide (CO_2) injection. This method of enhanced oil recovery has two main advantages including the ability to increase oil recovery significantly, and the ability to store a large portion of the CO_2injected downhole. This research aims to perform a detailed review of the difference in mechanism between immiscible and miscible CO_2 injection in shale reservoirs to illustrate the mechanisms, benefits of each method, and the optimum conditions at which each should be applied to reach the maximum oil recovery possible, and the largest storage potential. The main difference in mechanism between miscible and immiscible CO_2 injection comes in the interfacial tension (IFT). In miscible CO_2, the IFT is reduced to zero, however, unlike common belief, immiscible CO_2 injection will also have an impact on IFT reduction due to CO_2solubility in the oil. Miscible CO_2 usually is applied best with light and intermediate hydrocarbons due to the vaporization-condensation mechanism, while immiscible CO_2 injection is applied best with heavy oils due to the oil swelling mechanism. In terms of CO_2 storage, it was found that it is a function of both the fluid injected, and the rock itself. Increasing the injection pressure will result in a higher adsorption which in turn will result in a larger storage potential. If there is a large concentration of minerals that can interact with the CO_2 however, the storage potential is then impacted, either negatively or positively depending on the situation. This research aims to provide a detailed review of miscible and immiscible CO_2 injection in shale reservoirs and determine the best application conditions for each to reach the highest recovery and storage potentials.

An Experimental Study of the Factors Impacting Asphaltene Stability in Crude Oil during Carbon Dioxide Injection in Nanopores

Authors:
Fakher, S., Missouri University of Science and Technology
Imqam, A., Missouri University of Science and Technology

Carbon dioxide (CO_2) injection has been applied recently in unconventional shale reservoirs to improve oil recovery from the nano-pores. During CO_2 injection, asphaltene may begin to precipitate, and eventually deposit into the nanopores of the shale, which will result in formation damage and pore plugging. This research investigates asphaltene pore plugging in nano-pores of shale reservoirs during CO_2 injection, and its impact on formation damage and oil recovery. A novel high pressure high temperature experimental setup with multiple nano-filter membranes was designed and constructed to study asphaltene impact on oil recovery. The effect of varying CO_2 injection pressure, temperature, oil viscosity, CO_2 soaking time, and pore size was studied. Asphaltene pore plugging was observed in all experiments conducted. Increasing the pore size from 10 to 100 nm resulted in a significant decrease in oil recovery, which reached 11.2% using 10 nm and 23% using the 100 nm. Heterogeneity increased the difficulty of the oil to propagate thorough the nano-pores. This resulted in an extremely low oil recovery, reaching up to 3.44% in the most severe heterogeneity design. The soaking time is the time at which the CO_2 is allowed to interact with the oil. Increasing this interaction time resulted in an early high oil recovery, which was then reduced due to higher asphaltene pore plugging, which reached 42% for the 2 hour soaking time. Increasing CO_2 injection pressure resulted in an increase in oil recovery, with the 700 psi pressure resulting in an oil recovery of 43% of the pores, whereas the 200 psi resulted in 8.21% pore plugging. The temperature increase resulted in a significant increase in oil production rate, however asphaltene pore plugging increased with time. Higher oil viscosity had higher initial asphaltene concentration and also much lower oil mobility which resulted in extremely low oil recovery.

An Overview on the Ieaghg Technical Programme: CO2 Capture Technologies for the Power and Industrial Sectors, Their Integration and Potential to Reduce Costs

Authors:
Garcia, M., IEAGHG

The first generation of carbon capture technologies has been proven to be feasible and the second generation seems to offer further enhancements. While models usually include 90% of capture rate, that can be predictably increased. In the power sector, the main focus has been the reduction in cost and energy penalty. However, many discussions on the value of CCS beyond LCOE are still ongoing as part of the future complex electricity grid in different locations. Moreover, the water consumption has been highlighted as one of the concerns for water-stressed regions and a further assessment was needed.

In the industrial sector, carbon capture technologies can tackle down process emissions which could not be reduced otherwise. However, the preferred configuration is site-specific and a tailored analysis is needed for each case of study.

To solve those concerns, IEAGHG has commissioned seven technical studies and reviews in the period 2017-2019. In this presentation, the results from the IEAGHG technical studies will be discussed, including an overview of the benchmark and emerging CO_2 capture technologies. The techno-economic analyses have shown that emerging technologies might be promising, perhaps less detailed cost assessments for low TRL technologies are published in the literature due to the lack of demonstration projects. Flexibility is key for the integration of CCS in the electricity grid and the costs and water consumption in power plants are strongly dependent on the location, due to local conditions and regulations. Moreover, it is techno-economically feasible to increase the capture rate. Those parameters will impact on the decisions for integrating and designing CCS systems. In the industrial sector, energy/steam production, and access and characteristics of the electricity grid are decisive parameters to select the CO_2 capture system configuration.

Analytical Solutions of Leakage through the Caprock in Carbon Sequestration in a Closed-Boundary Aquifer System

Authors:
Mu, L., China University of Petroleum - Beijing
Liao, X., China University of Petroleum - Beijing
Zhao, X., China University of Petroleum - Beijing
Zou, J., China University of Petroleum - Beijing
Chu, H., China University of Petroleum - Beijing
Li, R., China University of Petroleum - Beijing
Zhang, J., CNPC Engineering Technology R&D Company Limited
Deep saline aquifer is deemed as an effective and promising site for carbon sequestration. A geological storage site must be operated safely; however, CO_2may leak through several pathways, openings in the caprock being one of them. Many efforts have been made to monitor the leakage.

This study presents a new analytical method obtaining the pressure change in the monitoring well caused by the leakage through the openings in the caprock to detect the leakage and evaluate the leakage rate. The model is based on the assumptions that the leakage is through the opening in the caprock and the outer boundary of the system is closed. The pressure changes are determined with Laplace transform and superposition principle and then the leakage rate is determined.

The comparison shows that the analytical and numerical solutions are in good agreement. Through the analyses, it can be concluded that the height and permeability of the monitoring aquifer, the location of the monitoring well and the location and the permeability of the opening in the caprock do have some impacts on the leakage. When the opening is close to the injection well or the leakage permeability is high, the leakage rate will increase significantly. The influences of the monitoring aquifer above the sequestration aquifer reflect on its height and permeability. With the increase of the height and the permeability leading to a lager pressure difference, the leakage rate will be larger. The leakage rate is unaffected by the location of the monitoring well. However, decreasing the distance between the injection well and the monitoring well causes the increase of the monitoring pressure. The analytical model in this paper provides a quick and accurate evaluation for the leakage through the caprock in a closed-boundary aquifer system.

Bringing It All Together: Innovative Public Policy and Technology for Carbon Purification Catalyzes Power Production, Enhanced Oil Recovery, and Geological Sequestration in Depleted Oil Reservoirs: An Oklahoma Example

Authors:
Tormey, D., Catalyst Environmental Solutions Corporation
Schuller, R., PLASMA Power Products

The Oklahoma Conservation Commission developed the Carbon Program to provide state-verified carbon credits that incentivizes geologic carbon sequestration, including enhanced oil recovery by carbon dioxide (CO_2) injection underground. The United States (US) Department of Energyâ€™s carbon capture, utilization, and storage initiative further incentivizes public-private collaboration.

Depleted oil and gas reservoirs are ideal candidates for carbon sequestration, combining capacity and trapping characteristics. In moving to full geological sequestration, injection of CO_2 enhances oil recovery (EOR) in some formations. Oklahoma has opportunities that would increase oil production levels by over 20 percent. However, a significant hurdle is the availability and transport networks for CO_2. Once the infrastructure is in place for EOR, it remains as the depleted field transitions to full geological sequestration.

In addition to current infrastructure limitations, another impediment to wide-scale implementation of EOR is the supply of pure source CO_2. PLASMA is developing commercial carbon capture applications for both existing and greenfield coal-fired power plants and subsequent sequestration projects in the United States through their patented proprietary in-combustion technology for coal or coke combustion, wherein the captured carbon dioxide is of superior quality suitable for use in EOR operations. An independent analysis determined that PLASMAâ€™s in-combustion capture is significantly less expensive than both pre-combustion and post-combustion capture techniques. Furthermore, the PLASMA technology results in higher recovery efficiencies, at 100 % CO_2 capture, than lower post-combustion capture recovery. PLASMAâ€™s technology captures CO_2 with ultra-low sulfur oxides (SOx) and nitric oxides (NOx) content, thereby reducing partial pressure downhole requirements needed for EOR in mature oil producing reservoirs.

We have facilitated workshops among policy makers, power producers, and oil and gas companies in Oklahoma to bring it all together, balancing economic benefits of EOR and verified carbon credits against infrastructure cost.

Carbon Dioxide Capture By Gas Hydrate Based Technology

Authors:
Pandey, G., National University of Singapore
Kumar, R., Indian Institute of Technology Madras
Kushwaha, O. S., Indian Institute of Technology Madras
Bhattacharjee, G., National University of Singapore
Sangwai, J., Indian Institute of Technology Madras

With the increase in industrial as well as urban growth, there has been a momentous escalation in the emissions of greenhouse gases, in which carbon dioxide is foremost in the list. Although, carbon dioxide can be recycled back into the bio-system naturally, there is massive gap in the natural process of carbon sequestration and the rate of emissions. Additionally, a direct correlation between the carbon dioxide emissions and global warming has been established. It has been estimated that by the end of this century, the earthâ€™s temperature would rise by a margin of 4^0C, which will have a catastrophic impact on the entire biosphere. One option could be to reduce the carbon dioxide emissions from use of fossil fuel and identify an ecologically feasible method for clean energy generation. The other option which is more practical is to sequester the emitted carbon dioxide from the anthropogenic source. Storing of carbon dioxide as hydrates in permafrost and oceanic region over the world is an innovative technique of carbon capture and storage. Temperature and pressure condition which exist under the sea bed is conducive for hydrate growth and gaseous CO_2which is otherwise mobile can be sequestered in solid hydrate form for ever. We carried out the experimental studies in laboratory setup consisting of temperature controlled high pressure autoclave to study the thermodynamic and kinetic feasibility. One of the objective was to enhance the hydrate formation kinetics by supplying a suitable additive with CO_2. The kinetics of the carbon dioxide capture can be enhanced by using suitable additives and one need not wait for geological time scale for CO_2 hydrate formation and sequestration.

Carbon Dioxide Decomposition Using Oxygen Deficient Metal Oxides

Authors:
Park, C. Y., Korea Institute of Energy Research
Kim, S. H., Korea Institute of Energy Research
Jang, J. T., Korea Institute of Energy Research
Sim, J., Korea Institute of Energy Research
Lee, J. H., Korea University
Nam, S. C., Korea Institute of Energy Research
It is generally believed that the climate change is mainly attributed to the enormous emission of greenhouse gas in a short period of time. Although many efforts have been devoted for the treatment of CO_2, there is almost no realistic solution except the injection to the underground. The storing into the underground also has several problems including leaks, earthquakes, and the costs. Although Tamaura et. al reported that carbon dioxide could be reduced ~100% to carbon at 290 °C, it was the results obtained only in a lab-scale batch type reactor. For the practical applications, the study of decomposition in the continuous flow of carbon dioxide is essential. In this report, we demonstrate the decomposition results of CO_2 using spinel-type Ni ferrites and strontium and iron-based non-perovskites. Based on the suggested reaction mechanism, the experimental conditions and the catalysts were selected and tested in a continuous CO_2 decomposition reactor. We also studied the structural behaviors of the catalysts before and after the measurements.

Demonstration of Hydrogen Production Process Using Hyperthermophilic Archaebacteria Thermococcus Oonnurineous NA1

Authors:
KIM, H., Ajou University

The increasing demand for clean energy, global concern about climate change and problem related to fine particulate matter have resulted in an increased global willingness to leas hydrogen economy as a potential long-term solution to the environmental crisis. Although most hydrogen is produced by the steam reforming of natural gas, it can be produced from various feedstocks and production pathways. Hydrogen can be produced from the gas containing carbon monoxide based on water gas shift reaction in thermochemical or biological process. Biological water gas shift processes, while slower than chemical reactions, have a number of advantages such as higher yields, requirement of minimum energy due to lower operating temperatures and pressures, and lower cost. By-product gas in steel manufacturing plant or syngas in coal gasification plant which contains carbon monoxide can be converted to hydrogen gas using water gas shift process. In this study, hydrogen production process design and technological assessment have been conducted based on biological water gas shift process using Thermococcus onnurineus NA1 which is microorganism in deep-sea hydrothermal vent areas. Syngas derived from coal gasification plant could be an attractive source of CO for hydrogen production by biological water gas shift reaction. Demonstration scale hydrogen production process is designed and constructed with the capacity of 500 Nm^3/h syngas from coal gasification plant. As a result of process design, CO conversion efficiency of the process is 94.4% and 0.98 ton/d of hydrogen that can be utilized to run around 2,000 hydrogen fuel cell vehicles can be produced.

Keywords: Hydrogen, Biological water gas shift, Thermococcus onnurineus NA1, Syngas

Direct Thickening of Supercritical Carbon Dioxide Using CO2-Soluble Polymer

Authors:
Al Yousef, Z., Saudi Aramco
Swaie, O., Saudi Aramco
AlAbdulwahab, A., Saudi Aramco
Kokal, S., Saudi Aramco

Two major applications of injecting dense carbon dioxide (CO2) into the petroleum reservoirs are enhanced oil recovery and sequester CO_2 underground. For enhanced oil recovery applications, the low viscosity of injected CO2 compared with the reservoir fluids causes the fingering of CO2, which may lead to bypassing huge amount of oil, early breakthrough of CO2, and increasing the gas to oil ratio. The use of direct thickeners, such as polymers, is one of the techniques used to increase the CO2 viscosity. Nevertheless, the solubility of polymers in CO2 and the high cost of soluble polymers are the main challenges. In this study, a novel, soluble, and cost effective thickener is proposed to directly increase the CO2 viscosity.

A PVT swelling apparatus is used to assess the compatibility and solubility of the thickener in dense CO2. Also, a custom designed apparatus is used to measure the viscosity of the mixture of thickener and dense CO2 at reservoir conditions. The assessment is conducted at different pressures, temperatures and thickener concentrations.

The results reveal that there must be a minimum pressure at which the thickener dissolves in the dense CO2. The minimum solubility pressure is found to be around 1500 psi at 25 °C, and it increases with temperature. The results also demonstrate that the viscosity increases as the pressure increases. The increase of thickener concentration has a significant impact on CO2 viscosity. The measured CO2 viscosity at 0.30 wt% thickener concentration, 3000 psi, and 50°C is 3.88 cP. The addition of 0.60 wt% at the same experimental conditions results in a viscosity of 36.04 cP. This enhancement ranges between 65 and 1200 fold over pure sc-CO2 viscosity.

The injection of viscous CO2 has the potential to mitigate the gravity override and solve the CO2 mobility challenges.

Effect of Water-alternating-gas Injection on Gas and Water Production Control in Carbonate Reservoirs

Chuang Zhao[a], Zhaojie Song[a,*], Yang Yao[a], Yunbo Li[b], Qiang Liu[a], Mei Qi[b], Jirui Hou[a], Likun Xu[b], Bin Han[b], Liguang Song[a,†]

[a] Unconventional Petroleum Research Institute, China University of Petroleum, Beijing

[b] Research Institute of Petroleum Exploration & Development, CNPC, China;

Abstract

The viscous fingering during continuous gas flooding and water channeling during water flooding will reduce the sweep efficiency and ultimately affect the displacement efficiency. In this paper, extensive experiments with different injection modes were conducted to study their effect on production performance.

Experimentally, based on the geological characteristics and development conditions of the target carbonate reservoir, a laboratory physical model was established to study the migration characteristics and occurrence state of the injected gas and water. The effect of different injection modes, including continuous gas flooding, water-alternating-gas (WAG) flooding and water flooding, on water and gas production and oil-displacement efficiency was investigated. Experimental results indicated that the rising of water cut of WAG flooding was much lower than that of water flooding, presenting a good effect on delaying water production. WAG injection showed a relatively plateau period of gas-oil ratio, indicating injected water delayed gas channeling. WAG flooding achieved the highest oil recovery, followed by continuous gas flooding and water flooding.

This study provided an improved understanding of gas and water channeling behavior, and proposed schemes for delaying gas channeling and improving gas flooding.

Key words: gas flooding; water flooding; gas and water production; water-alternating-gas injection; physical simulation.

* Corresponding author, songz@cup.edu.cn (Z. Song).

† Presenting author.

1. Introduction

The target reservoir is a heterogeneous reservoir, and water injection or gas injection is generally used for oil production. There are a lack of natural energy, slow effect, high water cut in the middle and late stages, oil production declining and other issues for conventional water injection [1-3]. Gas (CO_2) flooding is one of the main technologies for enhancing oil recovery. Its unique mechanism of enhanced oil recovery includes viscosity reduction by dissolution, light fraction extraction and interfacial tension reduction. Besides of improving oil recovery, it can also play a role in reducing greenhouse gases and achieve a win-win situation for oil displacement and storage. But gas injection has problems such as premature gas timing and serious gas channeling [4]. Water-alternating-gas (WAG) flooding, as a gas injection method alternately with gas flooding and water flooding, combines the advantages of gas flooding to extract light components, reduce oil viscosity and interfacial tension, and improve the microscopic sweep efficiency of water flooding [5]. It can compensate for the problem of insufficient supply and gas channeling to effectively improve oil recovery.

Based on extensively researches and applications of gas flooding technology, and combined with the target reservoir geology and development conditions, we established an indoor experimental physical model and studied the oil recovery of WAG technology, as well as the behavior of water cut and gas-oil ratio. Comparing the prodcution performance of three different injection methods, the influence of different injection methods on oil recovery was revealed [6-7].

2. Experiments

2.1 Materials and experimental set-up

The materials used in the experiment, including oil, water, gas and core samples, and the experimental stet-up were the same as that in the paper "Experimental Modeling of Gas Channeling for Water-alternating-gas flooding in High-temperature and High-pressure Reservoirs (CMTC-558864)".

2.2 Experimental procedures and schemes

2.2.1 Water-alternating-gas (WAG) flooding

An artificial carbonate core whose dimensions, permeability ratio, injected rate and back-pressure are given in Tab.1 was wrapped with a mixture of epoxy resin and curing agent.

Tab.1 Basic parameters of alternating injection of water and carbon dioxide

Core permeability (mD)	Back pressure (MPa)	gas-water volume ratio	Gas injection rate (ml/min)	Water injection speed (ml/min)
50/250	25	1：1	0.2	0.2

For coreflooding experiments, the sequence of experiments was as follows.

①Artificial carbonate cores coated with epoxy resin and curing agent were put into the core holder and then confining pressure was added. ②The core was vacuumed for 4-6 hours and saturated with formation water to calculate the permeability of water phase. ③Calculate oil saturation after the core was saturated with crude oil. ④The aging stage of the oil, which lasted 24 hours, was designed to be consistent with the form of crude oil that has been stored in the actual formation for countless years. ⑤ According to experimental requirements, different permeability ratio and injection rates were set. ⑥In the way of first gas injection and then water injection cycle alternating injection, the gas-water volume ratio was 1:1. ⑦The experiment was terminated when the producing gas-oil ratio reached 3000 mL/mL.

2.2.2 Water flooding and continuous gas flooding

Water flooding and continuous gas flooding experiments were conducted to compare with the result of WAG flooding. The temperature was set to 84 ℃, and the back pressure was set to 25MPa. The injection rate was 0.2 ml/min. The experiment was stopped until the water cut was greater than 98% or the gas-oil ratio was greater than 3000 mL/mL.

3. Results and discussion

3.1 Analysis of WAG displacement law

As shown in Fig. 1, the entire WAG process can be divided into four production stages. 0-0.1 PV is the first stage, mainly the process of supplying energy and gas dissolution accompanied by viscosity reduction, the oil recovery at this stage is 0.42%. 0.1-0.26 PV is the oil recovery stage without water, the oil produced in this process does not contain water, the oil recovery is 21.77%. The third stage (0.26-0.6 PV) is the stage of simultaneous production of oil, gas and water, also the stage with the highest oil recovery rate in the entire WAG process. Since the flow capacity of the gas is stronger than that of the water, the gas is produced preferentially over the water, and maintain fluctuations in a lower gas production range. When the injection volume reaches 0.307 PV, water production begins, and the water cut increases in a wave-like zigzag. At this stage, the interaction between gas and water effectively prolongs the time of water and gas channeling, and is the most productive and important oil recovery stage in the WAG process. The fourth stage is the gas channeling stage, the ability of gas and water to interact disappears. Gas-oil ratio and water cut rise sharply, no longer fluctuating. The oil recovery rate is weak and basically stabilizes.

The first stage: Mainly to supplement the formation energy and gas dissolution and viscosity reduction stage (t=0.1 PV, Ro=0.42%);

second stage: Waterless production stage (t=0.1-0.26 PV, Ro=21.77%);

The third stage: At the stage of simultaneous production of oil, gas and water, the water content increases in a wave-like tortuosity (t=0.26-0.6 PV, Ro=48.44%);

Fourth stage: Gas channeling（t=0.6-0.75 PV， Ro=1.25%）.

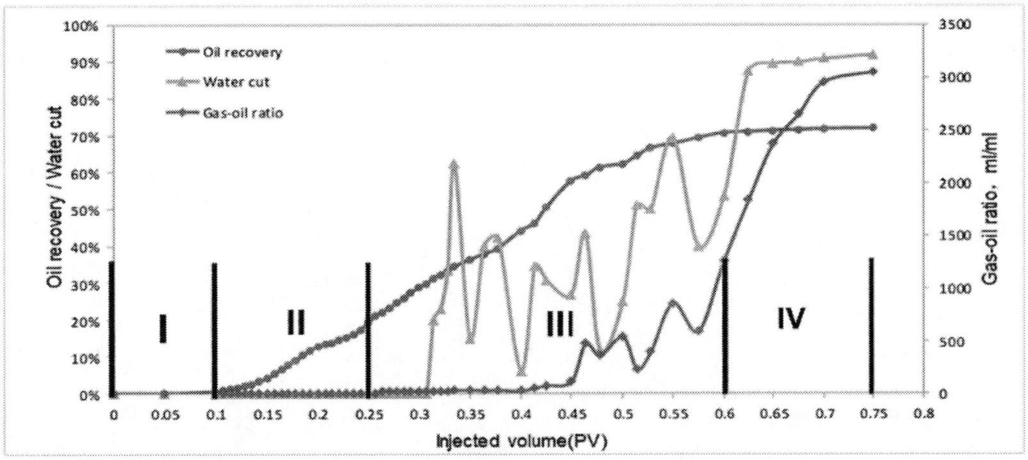

Fig. 1 Producing gas-oil ratio, water cut and oil recovery of continuous gas flooding in heterogeneous core sample (Q=0.2ml/min, vk=5)

3.2 Comparison of water cut changes

As shown in Figs. 2 and 3, WAG see water time (t=0.307 PV) is much later than continuous water flooding (t=0.093 PV), and the rate of water cut rises much less than continuous water flooding, the law of water rise rises from S type to concave type. It shows that the injection of gas plays an effective role in water control, which makes the oil recovery of WAG far higher than that of continuous water flooding.

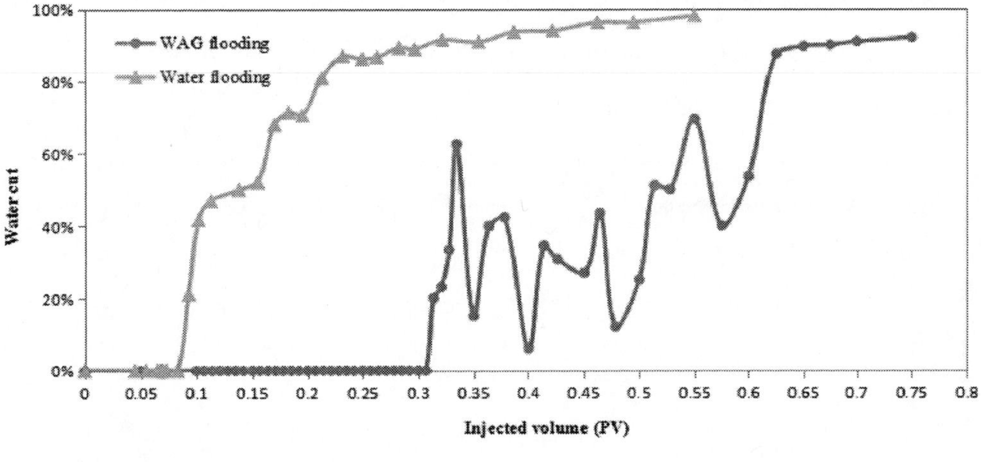

Fig. 2 Comparison of water cut of water flooding and WAG flooding

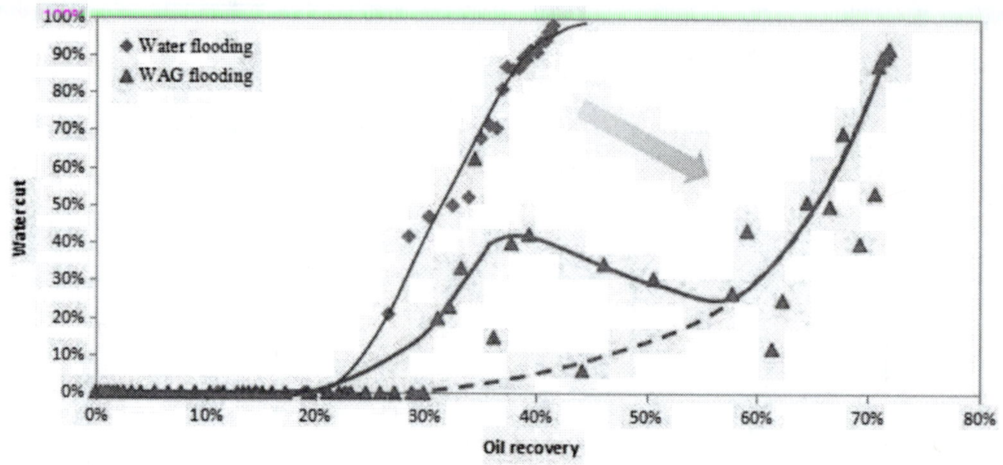

Fig. 3 Relationship of water cut and oil recovery for water flooding and WAG flooding

3.3 Comparison of gas-oil ratio

As can be seen from Fig. 4, the rising phase of the gas-oil ratio during the WAG process (Ro=60%) lags significantly behind the continuous gas flooding (Ro=40%). A large amount of crude oil was produced under low gas-to-oil ratio conditions, indicating that the injection of water played a role in delaying the gas. However, in the later stage, the formation of gas channeling and the loss of water to inhibit the ability of gas led to a sharp rise in gas-oil ratio. The increase in oil recovery rate tends to be flat, which also indicates that although the injection of water can delay the gas channeling, it cannot completely prevent the gas channeling.

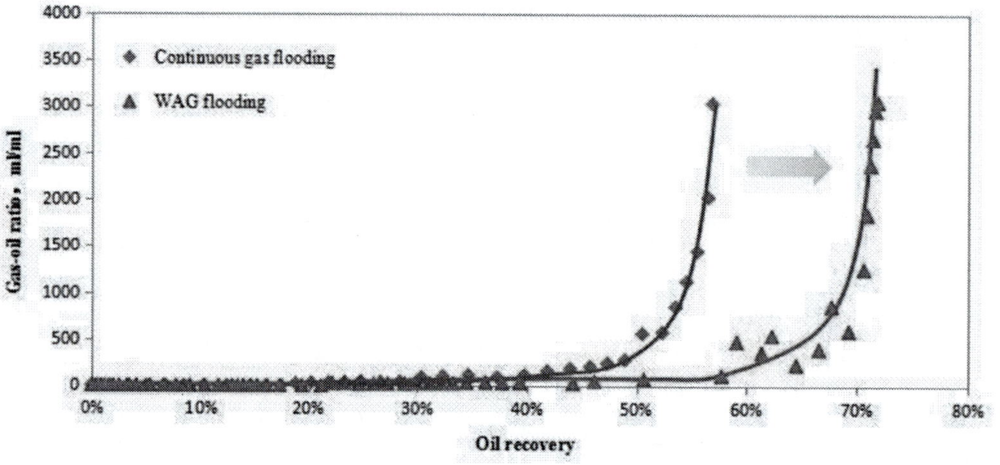

Fig. 4 Relationship of gas-oil ratio and oil recovery for continuous gas flooding and WAG flooding

3.4 Comparison of oil recovery

Comparing Figs. 1, 5, and 6, at the same pressure, temperature and injection rate, the gas flooding recovery is better than water flooding for cores with a permeability difference of 5, which can be increased by about 14 percentage points. The oil recovery rate of WAG is significantly better than gas flooding, which can increase by about 15 percentage points. In the process of WAG, the water cut first rises and then falls, showing a tendency of reciprocating fluctuations. After seeing the water, in the range of fluctuations in water cut, the rate of oil recovery increases rapidly, which is a high-yield period.

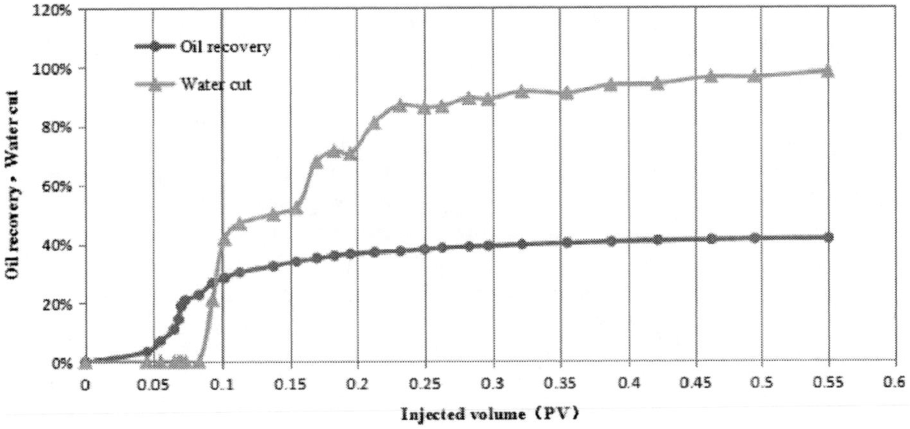

Fig. 5 Water cut and oil recovery of water flooding

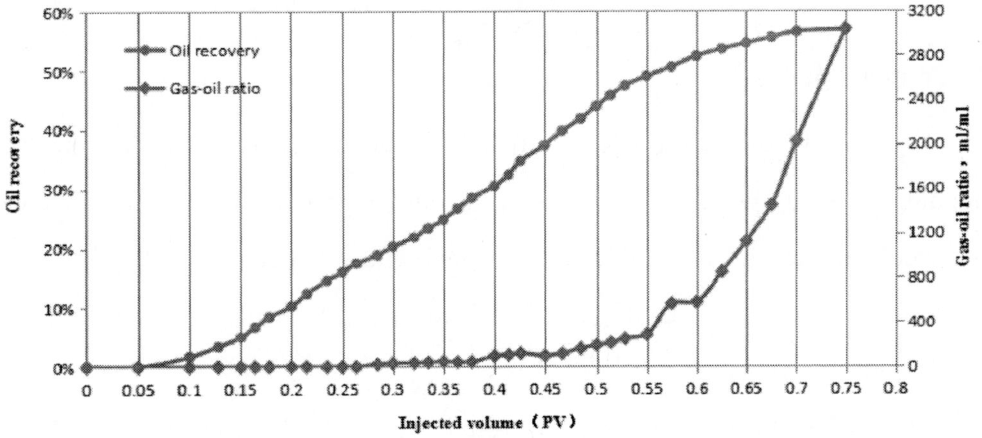

Fig. 6 Producing gas-oil ratio and oil recovery of continuous gas flooding

Comparing the production performance of three different injection methods (as shown in Fig. 7), the injected water has low compressibility and strong energy supplement effect, the effect of improving oil recovery in the early stage is obvious, and the displacement in the middle and late stages is obviously insufficient. In the mid-term of gas flooding, the effect of enhanced oil recovery is more obvious, but gas channeling will greatly reduce the oil recovery. The technology of WAG combines the advantages of both water flooding and gas flooding, and the oil recovery rate is the most obvious, with the highest degree of recovery.

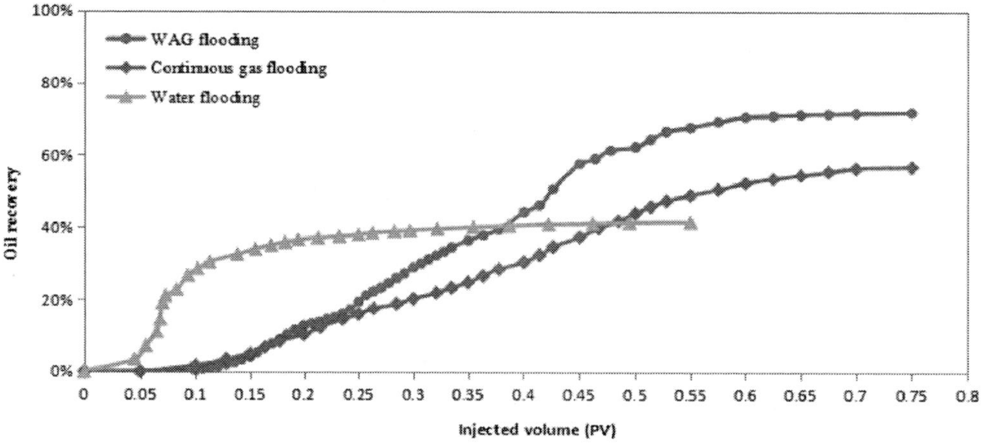

Fig. 7 Oil recovery of water flooding, continuous gas flooding and WAG flooding

4. Conclusions

(1) WAG flooding achieved a significantly higher oil recovery than continuous gas flooding and water flooding. Compared with continuous gas flooding, the oil recovery rate can be increased by about 15%. Compared with water flooding, the oil recovery rate can be increased by about 30%. WAG combined both the advantages of water flooding and gas flooding.

(2) Water-alternating-gas injection presented a good effect on water control. In the process of WAG, the water cut first rose and then fell, showing a tendency of

reciprocating fluctuations. After water breakthrough, in the period of water cut fluctuation, the oil recovery increment was highest and the duration was the longest.

(3) Gas-oil ratio increased sharply after 0.6 PV was injected for both continuous gas flooding and WAG flooding.

References

[1] YUAN Shiyi, WANG Qiang. New progress and prospect of oilfields development technologies in China [J]. Petroleum Exploration and Development, 2018, 45 (4) : 657-668.

[2] HU Wenrui, WEI Yi, BAO Jingwei. Development of the theory and technology for low permeability reservoirs in China [J]. Petroleum Exploration and Development, 2018, 45 (4) : 646-656.

[3] Li Shilun, ZHOU Shouxin, DU Jianfen, et al. Review and prospects for the development of EOR by gas injection at home and abroad [J]. Petroleum Geology and Recovery Efficiency, 2002, 24 (2) : 1-5.

[4] Li Zhenquan, YIN Yong, WANG Qiwei, et al. Development of the research on EOR mechanism by WAG [J]. Journal of Southwest Petroleum University (Science &Technology Edition) , 2007, 29 (2) : 22-26.

[5] ZHOU Xingze, LIAO Xinwei, ZHAO Xiaoliang, et al. Laboratory investigation of miscible CO_2 WAG process in extra-low permeability reservoir [J]. Journal of Shanxi University of Science & Technology (Natural Science Edition) , 2016, 34 (6) : 120-124.

[6] CHEN S, LI H, YANG D, et al. Optimal parametric design for water alternating-gas (WAG) process in a CO_2-miscible flooding reservoir [J]. Journal of Canadian Petroleum Technology, 2010, 49 (10): 75-82.

[7] NEZHAD SART, MOJARAD MR, OSKOUEI S J P, et al. Experimental study on applicability of water-alternating-CO_2 injection in the secondary and tertiary recovery [R]. SPE 103988, 2006.

Effectiveness of CO2 Foamed Acid for Stimulating Low Permeability Carbonate Formations

Authors:
AlTaq, A., Saudi Aramco
Al Yousef, Z., Saudi Aramco

Carbon dioxide (CO2) has so many applications in petroleum industry. In carbonate water injection wells, acid stimulation treatments are conducted as a common practice following drilling of the target zone, to remove mud-induced formation damage and improve formation permeability. Obviously, injected acids will preferentially go to the non/less damaged zones (high permeability) over the damaged zones (low permeability), which will lead to poor zonal coverage. Being more stable in high-permeability layers, CO2 foams can be used to provide effective acid diversion during acidizing treatments.

Four surfactants were used to evaluate the foam generation and rheological properties of different CO2 foamed acid systems at ambient and reservoir conditions, e.g., high-pressure, high-temperature (HPHT). Compatibility, thermal stability, bulk, and dynamic tests were conducted to evaluate the effectiveness of each acid system to generate stable and viscous foams. Dynamic foam viscosity measurements was conducted in a special foam rheology apparatus at HPHT. The foam was generated by injecting gas and surfactant containing acid solutions at different surfactant concentrations, acid concentrations, shear rates, injection qualities, and temperatures.

The results show that there is a direct correlation between surfactant concentration and foam strength. Also, the results indicate that as the acid concentration increased, the foam viscosity decreased. Foams exhibited shear thinning behavior, where foam viscosity decreased with increased shear rates. Also, the results show that the foam quality had a significant impact on foam stability and viscosity. For each surfactant-acid solution, there was a certain quality at which the highest stability and viscosity was attained. The results also report that the temperature has a negative impact on foam stability and viscosity, with the foam becoming less stable and weak with increased temperature.

This paper presents in detail the role of CO2 foamed acid on improving diversion stages in the stimulation treatments.

Effects of Electrolyte Environment on Carbon Dioxide Reduction

Authors:
Lucci, F., Lawrence Livermore National Lab
Baker, S., Lawrence Livermore National Lab
A promising method to convert carbon dioxide into value-added chemicals is the electrochemical reduction of carbon dioxide on copper-based catalysts. Copper based catalysts are known to produce small organic molecules including ethylene, methane, and higher hydrocarbons, but with low selectivity to a given product. Previous studies suggest that the product distribution is dependent on the environment surrounding the electrode, but the interactions between the electrolyte and catalysts are unknown. To understand these interactions, we systematically altered the composition and pH of the electrolyte to probe the catalytic activity using in situ techniques. We used the unique combination of Fourier-transform infrared spectroscopy (FT-IR) and Atomic Force Microscopy (AFM) to probe both the formation of surface intermediates and changes to the electrode structure as a function of electrode environment. A full understanding of the synergy between the electrode and the surrounding environment will allow for the rational design of both the catalysts and electrochemical reactors.

Empirical Correlation of Pressure Interval of Near-Miscible Flooding for Pure and Impure CO2 Streams

Authors:
Chen, H., China University of Petroleum
Zhang, X., State Key Laboratory of Offshore Oil Exploitation

It is believed that the boundaries of near-miscible pressure region are in the vicinity and closely related to minimum miscible pressure (MMP). In this paper, over 140 data sets belonging to experimental MMP values along with enough corresponding gas/oil compositional information from the literature were collected to construct a new correlation for both pure and impure CO_2. Parameters such as reservoir temperature, molecular weight of C_7^+, solution gas and injection gas contents, are selected as the key input values. It is found that the proposed models of MMP significantly outperform the existing 10 commonly used methods and provide the best reproduction of MMP data. According to the data of previous study, the lower and upper boundaries for the pressure interval is determined as 0.87 MMP and 1.07 MMP, respectively. Thus, a new empirical correlation for the prediction of near-miscible pressure region was proposed for the first time. It should be noted that the upper boundary of the pressure interval is higher than the conventional MMP due to the inappropriateness of the determination criterion. Results showed that the newly proposed model can simulate the pressure interval based on the correlations of linear fittings and non-linear fittings for both pure and impure CO_2, respectively.

Evaluating Carbon Dioxide Flow Regimes in Micro and Nano Pore Sizes during Carbon Dioxide Injection in Hydrocarbon Reservoirs and Their Impact on Oil Recovery

Authors:
Fakher, S., Missouri University of Science and Technology
Imqam, A., Missouri University of Science and Technology
Hydrocarbon reservoirs can vary significantly from one another, with pore sizes ranging from micro-sized in conventional reservoirs, and high permeability features such as natural fractures, up to nano-sized in unconventional reservoirs such as shale. When injecting carbon dioxide (CO_2) in different pore sizes, the flow regime will vary significantly. Understanding what flow regime will dominate under different thermodynamic conditions, and in different pore sizes is imperative in order to be able to select the proper flow equation and to maximize oil recovery potential, and CO_2 storage capacity. This research studies the different flow regimes that may occur in different pore sizes of hydrocarbon reservoirs, ranging from 0.2 nanometer, and up to micro-sized pores during CO_2 injection using mathematical simulation based on the definition of Knudsen Number for flow regimes. Following the mathematical simulation, experiments were conducted using composite filter membranes with 2.7 micro-meter, 100, 10, and 0.2 nano-meter pore sizes under different pressure and temperature conditions to study the impact of different flow regimes on oil recovery percentage and behavior during CO_2 injection . The mathematical modeling determined that for nanopores less than 10 nm in size, Diffusion was the main flow regimes, while at pore sizes less then 200 nanometers a transition flow regime was observed. In the micro-pore sizes the man flow regime was Viscous flow. After performing the oil recovery experiments, it was found that in the nanopore sized membranes, the CO_2 took a much longer time to breakthrough the filter membrane, and the oil recovery was considerably lower than that observed in the micro-sized membrane. Also, as the pressure increased, the flow became more turbulent, however, the oil recovery increased. Based on this research, it different flow regimes will occur in different pore size ranges based on which different oil production behaviors will be observed.

Evaluation and Optimization of Gas Volume on CO_2 Huff and Puff By Multiple Horizontal Wells in Fault-Block Reservoirs with Edge Aquifers

Zhixing Wang[1,2,3,4], Fenglan Zhao[2,3,5,6], Jirui Hou[1,2,3,4], Hongda Hao[2,3,5,7], Li Yan[6] and Shijun Huang[8], (1)Beijing Key Laboratory of Greenhouse Gas Sequestration and Oil Exploitation, Beijing, China, (2)CUP-CMG Joint Laboratory for Reservoir Simulation, China University of Petroleum (Beijing), Beijing City, China, (3)Basic Theory and Application Research Laboratory for Enhanced Oil Recovery in Low Permeability Oilfield, Key Laboratory of Tertiary Oil Recovery, CNPC, Beijing City, China, (4)Unconventional Petroleum Research Institute, China University of Petroleum, Beijing, China, (5)Key Laboratory for Greenhouse Gas Sequestration and Oil Exploitation in Beijing, Beijing City, China, (6)Research Institute of Enhanced Oil Recovery, China University of Petroleum (Beijing), Beijing, China, (7)China University of Petroleum (Beijing), Beijing, China, (8)China University of Petroleum(Beijing), Beijing, China

Abstract

Problems like small sweeping volume, short stimulating period and poor producing performance challenge the effectiveness of single horizontal well CO_2 huff 'n' puff in the fault-block reservoirs with edge water. The effect of CO_2 volume on multiple horizontal wells CO_2 huff 'n' puff has been studied in this paper to investigates the feasibility of CO_2 huff 'n' puff by horizontal well group. Studies on both isothermal property of CO_2 and the CO_2-crude oil properties under reservoir conditions offer the possible basis for this work. A 3D edge water model with three horizontal wells, which was designed and produced according to geological data, was used to simulate the process of CO_2 huff 'n' puff by horizontal well group. Subsequently, experiments were carried to investigate the impact of CO2 volume on water control and oil stimulation. Results show the CO_2 isothermal compressibility under certain pressure make it possible to inject gas with relatively low energy loss and to enlarge sweeping volume. Meanwhile, with more CO_2 mole composition in the crude oil, the mixture intends to be more flowable. Consequently, as the gas volume increases from 0.07PV to 0.13PV, the comprehensive water cut rate drop from 0.72% to 5.93%, and both oil and gas production increase accordingly. However, the gas use ratio dwindles as the volume exceeds 0.13PV.

Introduction

There is around 50%-60% original oil in place remained after the primary and secondary oil recoveries. The ever-increasing demand for crude oil has inspired the innovative technologies to stimulate hydrocarbon productions. Therefore, enhanced oil recovery (EOR) methods, involving the injection of CO_2, N_2, hydrocarbon gases, or chemicals into oil reservoirs, becomes increasingly important in petroleum industry. Among the EOR methods, CO_2 EOR has gained more momentum since CO_2 injection can effectively improve the oil recovery factor and reduce greenhouse gas emissions to some extent[1-4].

The oil recovery can be enhanced by CO_2 injection through different processes, like displacing oil left in the unswept areas by water displacement. In addition, considerable oil production can be enhanced through series of micro mechanisms, covering oil expansion, viscosity reduction, capillary effect mitigation due to CO_2-crude oil interfacial tension (IFT)[5-7], light hydrocarbons extraction[8], permeability alteration[9-10], mass transfer through diffusion and dispersion. Oil mobility can be drastically improved due to the significant reduction in oil viscosity once mixed with the injected CO_2[11,12]. Oil expansion due to CO_2 dissolution can enhance oil recovery by stripping oil drops off the grain surface and increasing oil volume over the residual saturation, contributing the formation of continuous oil phase[7]. Reduction in residual oil saturation can also be achieved by the extraction of light hydrocarbons upon exposing the oil to a sufficient CO_2-rich gas[8]. However, the extraction mechanism is susceptible to hydrocarbon compositions in the crude oil. Therefore, heavy crude oils are less affected by this mechanism compared to light crude oils[13]. The contribution of each aforementioned mechanism to oil recovery is governed by the pressure, temperature, and CO_2 solubility.

According to the reservoir characteristics of a domestic edge water effect, using the self-designed edge water action level well group , physical model for edge water reservoirs horizontal well group and the physical simulation experiment, study the effect of injection volume on symmetry stimulation effect and optimize it, combined with the nature of the reservoir under the condition of high temperature and high pressure CO2 and its interaction with formation crude features, analysis of gas injection for horizontal well group raised the influence degree of the effect of oil water control in CO2 stimulation, for the effect of injection volume optimization and analysis to provide basic data and theoretical basis

Experiment materials and process

The crude oil used in the experiment was obtained from a fault-block reservoir in the Jidong Oilfield, is the mixture of reservoir crude oil with kerosene. And the oil viscosity is 189mPa·s under the reservoir condition(60 ℃, 16.4 MPa). Brine of the edge aquifers was collected from the production well with a salinity of 1572mg/L, $NaHCO_3$ type. The CO_2 gas was from the Jinggao Gas limited company with a purity of 99.9%.

The CO_2 huff and puff by multiple horizontal wells schematically shown in **Figure 1** was utilized in the experiments. A radius core holder specifically designed to the 3D model size with an inner diameter of 45cm diameter and depth of 10cm was used as a radius core holder for the CO_2 huff and puff experiments (left of **Figure 1**). This special core holder could sustain pressures up to 20MPa and temperature to 120℃. There are two high-pressure pumps (Model 100DX, Teledyne Technologies) working separately to pressurize the CO_2 (1 in **Figure 1 right**) and control the flow rate of the edge aquifer (4 in **Figure 1.b**). Pressures in the radius core holder were measured using pressure transducers (JYB-KO-H, Beijing ColliHigh sensing technology co., LTD). The produced gas and liquid were separated and measured by gas flowmeters (LF420-S, Laifeng Scientific Technology co., LTD) and graduated test tube.

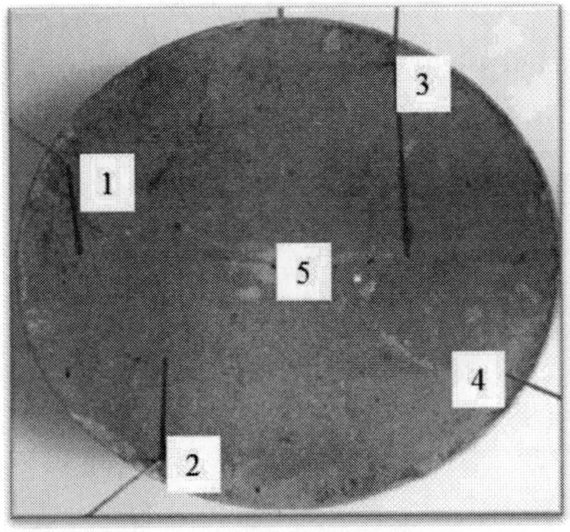

Figure 1 Radius Coreholder (left) and 3D model (right) for CO_2 huff and puff by multiple horizontal wells (red lines indicates the horizontal length and orientation of horizontal wells)

Experiments of CO_2 huff and puff by multiple horizontal wells of CO_2 volumes were performed in the laboratory using the experimental apparatus. The experimental procedures were as follows:

(1) Each 3D model was made from outcrop sand with 5 wells pre-buried according to the distribution patterns in Fig.1. The 3D model was evacuated for more than 4 h before saturated with brine, and the pore volumes and permeabilities were measured.

(2) The initial oil saturations and irreducible water saturations were calculated by the crude oil at a rate of 0.1 mL/min until the water production ceased. And the model was aged 48h at 60℃.

(3) Edge water flooding was initiated at back pressure of 7.5 MPa till the water-cut of one well reaches 98%, and the water flooding recovery was calculated.

(4) The high-pressure pump was initiated to inject CO_2 till the pressure in Well 5 was 3MPa, 5MPa, 7.5MPa and 10MPa.

(5) The soak time last 24h before all the horizontal wells recovered to production under 1 MPa BPR, and so did the edge aquifer flood. When water-cut of one horizontal well reaches 98%, flooding work ceased.

Table 1 Radius model parameters of CO_2 huff and puff cases

Experiment NO.	Pressure, MPa	bulk volume, cm^3	pore volume, cm^3	porosity, %	oil saturation, %
1	3	5652	842	14.89	65.32
2	5	5652	839	14.84	66.61
3	7.5	5652	810	14.33	64.56
4	10	5652	845	14.95	68.69

As introduced in the previous section, the CO_2 volume can be calculated by the density changes in a constant volume cylinder with a movable piston before and after injection at the constant temperature. CO_2 equation of state (EOS) in the density form is below,

$$PM_{CO_2} = z\rho_{CO_2}RT \tag{1}$$

The density difference before and after CO_2 can be described,

$$\Delta\rho = \rho_1 - \rho_2 = \left(\frac{P_1}{z_1} - \frac{P_2}{z_2}\right)\frac{M_{CO_2}}{RT} \tag{2}$$

The cylinder volume is 1000 cm^3 under this experiment condition, then the mass of the injected CO_2 can be calculated. Additionally, the volume under standard and experimental condition can be calculated by the CO_2 EOS in mass form (Eq. 3). Tab.3 summarized the gas injection volume under

$$PV = z\frac{m_{CO_2}}{M_{CO_2}}RT \tag{3}$$

Table 2 CO_2 injection volumes under different conditions by EOS calculation

Experiment NO.	Pressure, MPa	CO_2 mass, g	volume, PV (333.15K, injection pressure)	volume, PV (2933.15K, 0.1MPa)
1	3	3.294	0.07	2.1987
2	5	8.420	0.10	5.1233
3	7.5	18.030	0.13	9.7194
4	10	33.431	0.14	13.8035

Results and discussion

CO_2 compressibility under experimental conditions

Since the gas volume was measured according to the pressure value within the radius model. Under the reservoir condition, CO_2 was squeezed into the target zone by surface compressor, and gas compressibility coefficient was the key parameter to assess the injectivity and efficiency.CO_2 isothermal compressibility under different injecting pressure was evaluated (**Figure 2**). CO_2 compressibility decreased drastically with the elevating pressure and increased slightly near 8.5MPa. Namely, CO_2 was least compressed at pressure ranges of 6.5-8.5MPa at which CO_2 phase transformed from gas to supercritical state. During the injection process, CO_2 was easily compressed at gas phase, consuming large portion of energy to compress gas rather than to pump into the formation. Therefore, elevating injection pressure would achieve an economical energy efficiency for an expanded seeping volume.

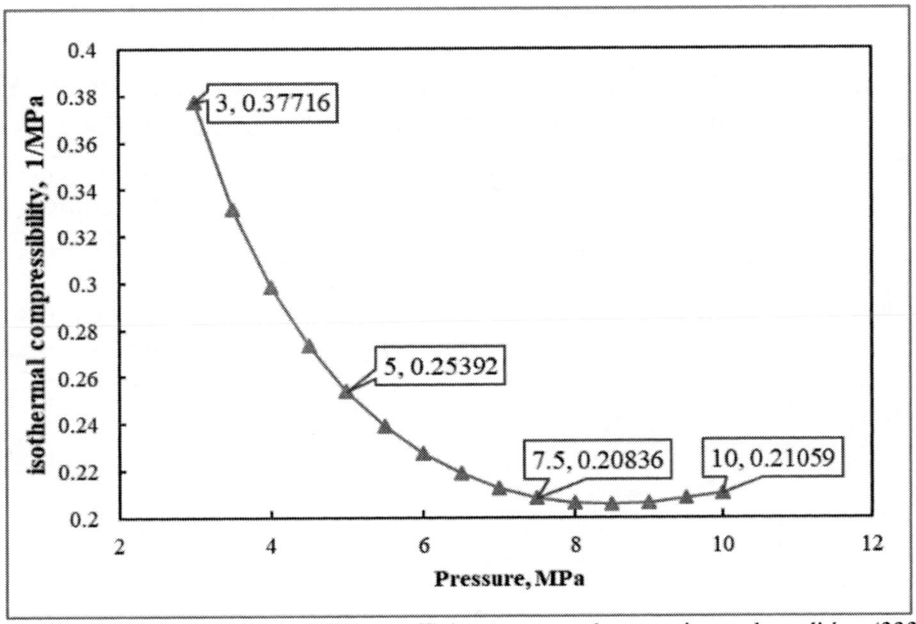

Figure 2 CO_2 isothermal compressibility coefficient curve under experimental condition (333.15K)

Oil properties alteration by CO2 solubility

The dissolution of CO_2 into crude oil caused oil expansion, mobility improvement and interfacial tension reduction, lowing the oil saturation in the formation. Under formation condition, oil volume expanded with the rise in CO_2 mole fraction. The volume increased 8% as the CO_2 mole fraction properties elevated to 30%. Therefore, the expansion of discontinuous oil droplets makes it possible to merge with the other isolated droplets, forming a continuous phase. Meanwhile, saturation pressure of the mixture increased to 9.38MPa, 2.13MPa higher than the original oil. The injection of CO_2 supplied the depleted formation energy. Consequently, gas oil ratio elevated to 35.5 A m^3/m^3 as large amount of gas dissolute as formation pressure depleted under saturation pressure, and "foamy oil" has a better mobility compared to the dead oil with lower viscosity, contributing oil flow into the wellbore.

Table 3 Oil properties mixed with different CO2 mole fractions

CO_2 mole fraction, %	B_o, dimensionless	P_b, MPa	GOR, m^3/m^3
0	1	7.189	0
5	1.0138	7.604	5.9
10	1.0277	7.995	11.8
20	1.0555	8.714	23.7
30	1.0835	9.389	35.5

CO_2 diffusion properties in crude oil

For CO_2 distribution in porous media, certain measured data were necessary to evaluate the CO_2 sweeping efficiency in the model. Pressure diffusion and molecular diffusion are the distinct mechanisms that affect the trend of CO_2 variation. Molecular diffusion is the tendency to mix due to concentration gradient. Pressure diffusion (gravitational segregation) the separation by pressure gradient. In our experiment, pressure was the indirect parameter to quantify the CO_2 injection volume, therefore, the two main mechanisms should not be neglected. Both CO_2 diffusion coefficient in the oil sample and CO_2 density under experimental condition various pressure were presented (**Figure 3**). Since the temperature was constant under experimental process, diffusion coefficient of CO_2 in the crude oil in various pressures condition was the key parameter to evaluate the CO_2 dispersion in the radius model during the soaking period. Results showed that the coefficient increased from $2.9927\times10^{-8}m^2/s$ to $4.8890\times10^{-8}m^2/s$, a near linear increase before the CO_2 critical state, and the tendency became debated after the critical point since the coefficient reduced to $4.7676\times10^{-8}m^2/s$ under 10 MPa condition. The round pot in Fig.2 depicted the tendency of CO_2density under experimental conditions. The oil and water densities are 0.837 $g/cm^3$and 1.15 g/cm^3 respectively. While the density of gaseous CO_2 is 0.0535 g/cm^3 at 3MPa, much lower than the formation liquids. The huge density gap leads to gravitational overlapping, contributing a gas distribution in the upper part of the model. When the difference narrows down to relatively high density like 0.1728g/cm^3 at 7.5MPa, the gas profile overrides the water swept area and expands oil by solution.

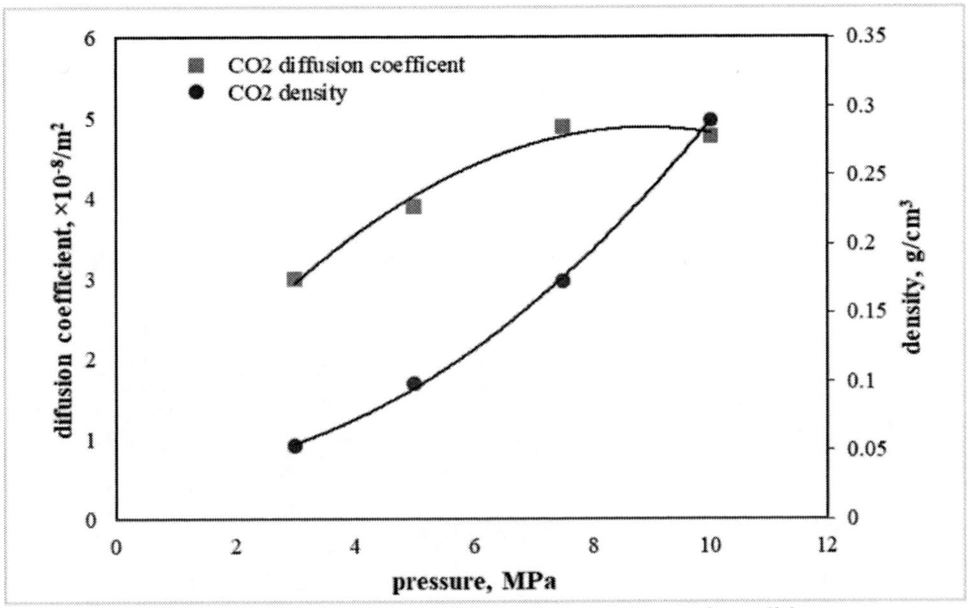

Figure 3 diffusion data of CO_2 under experimental condition
(diffusion coefficient were measured by pressure decay method and CO_2 density data were collected from NIST, a standard reference database)

Oil stimulation results and gas utility

Oil recovery results enhanced by CO_2 huff and puff were the comprehensive consequence of CO_2 oil interactions. **Figure 4** demonstrates the results of enhanced oil recovery by CO_2 huff and puff under different volumes. The stimulated oil increased in accordance with the enlarged CO_2 injection volume, as the overall oil recovery was 5.87% in 2.1987 SCPV and 22.36% in 13.8035 SCPV. More CO_2 meant a higher oil recovery. From the relation between oil stimulation and distance from edge aquifer in a single horizontal well, in the 2.1987 SCPV case, oil recovery in horizontal wells #1 and #2 was 2.60% and 2.02% respectively, higher than that in #3 which was only 1.25%, which was near the edge aquifer. And results of the other three experiments demonstrated the same oil recovery outcomes. For the horizontal well #3, which was located near the edge aquifer, much of the oil was flooded during the initial water flooding stage, then oil saturation was lower than the other two wells. Therefore, increasing gas injection volume can significantly increase the oil recovery. And the stimulation effect of CO_2 huff and puff mainly comes from horizontal wells with high oil saturation far away from the edge water.

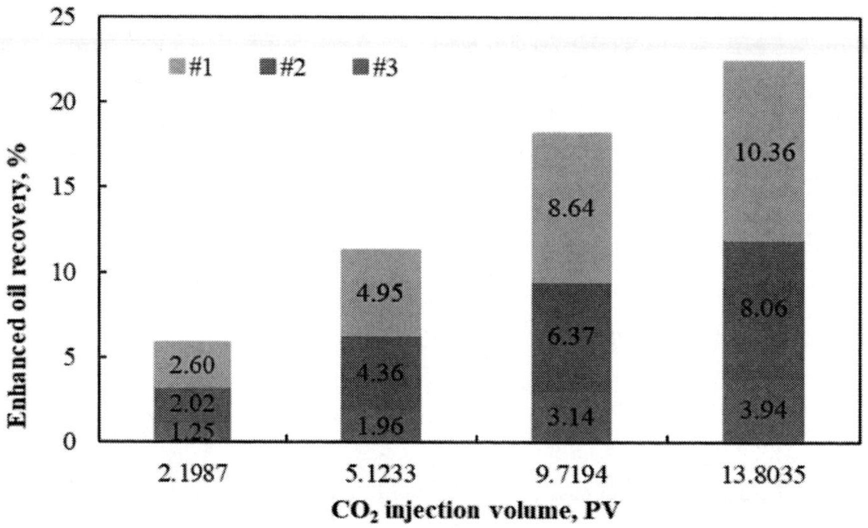

Figure 4 Enhanced oil recovery in single well of different CO_2 volumes
(gas volume in standard condition, corresponding to experimental condition 3MPa, 5MPa, 7.5MPa and 10MPa)

From the aspects of the enhanced oil recovery and gas utility (**Figure 5**), result of 9.37 scPV was 6.87% higher than that of 5.21PV, only 4.21% lower than that of 13.08 PV. Hence, the oil increase became less significant after 7.5MPa injection volume, and enlarging CO_2 injection volume would not stimulate oil drastically. Since small gas volume means light mass, the gas utility ratio, increased oil mass per 1g CO_2, decreased with the rising CO_2 volume. However, the gas utility ratio corresponding to 0.13PV under 7.5 MPa turned less obvious. In addition, free gas from production curves in Fig.3 also explained the shrank oil stimulation by unit mass CO_2 since the accumulated gas volume surged to 8000 cm^3 in 13.80 scPV.

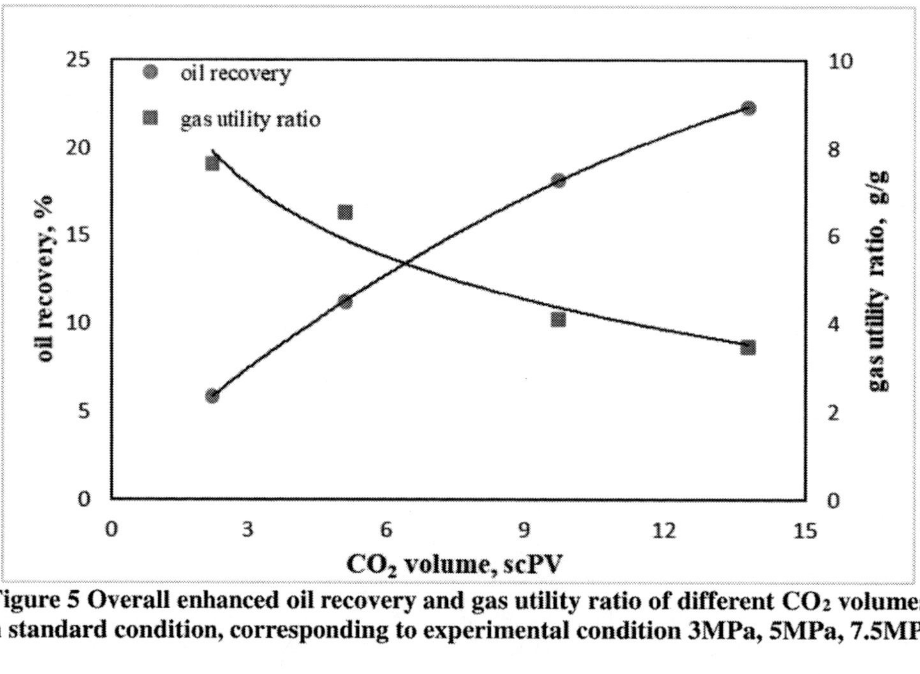

Figure 5 Overall enhanced oil recovery and gas utility ratio of different CO_2 volumes
(gas volume in standard condition, corresponding to experimental condition 3MPa, 5MPa, 7.5MPa and 10MPa)

Conclusion

From the results of injection volume optimization of CO_2 huff and puff from three horizontal wells in the radius model. Several conclusions has been reached as follows:

(1) The CO_2 huff and puff in the fault-block reservoir model with edge aquifer proved to be effective in water cresting control and oil stimulation. As injection volume of CO_2 was elevated, both the reduced water cut and enhanced oil production were observed, with substantial increase in free gas production. The injection volume needs to be optimized to alleviate the contradiction among water control, oil enhancement and high gas production.

(2) Based on the properties of CO_2 gas and CO_2-formation crude oil under high pressure conditions, the isothermal compression coefficient of CO_2 is the lowest under the experiment condition (60℃, 7.5MPa), corresponding to 9.72PV injection volume. The CO_2 gas can be squeezed into the formation, and the fluidity of crude oil increases with significant expansion with elevated CO_2 mole fraction.

(3) Under the experimental conditions of this study, results of water control and oil stimulation in 0.13 PV CO_2 was much better than that of 0.10PV, and similar to that of 0.14PV with much less gas production. And the optimal injection volume of CO_2 huff and puff in horizontal well group was 0.13PV.

Reference

1. Aycaguer A C , Lev-On M , Winer A M . Reducing Carbon Dioxide Emissions with Enhanced Oil Recovery Projects: A Life Cycle Assessment Approach[J]. Energy & Fuels, 2001, 15(2):303-308.
2. Khoo H H , Tan R B H . Life cycle investigation of CO2 recovery and sequestration[J]. Environmental Science & Technology, 2006, 40(12):4016-24.
3. Nobakht M , Moghadam S , Gu Y . Mutual interactions between crude oil and CO2 under different pressures[J]. Fluid Phase Equilibria, 2008, 265(1-2):94-103.
4. Abedini A , Torabi F . On the CO2 storage potential of cyclic CO2 injection process for enhanced oil recovery[J]. Fuel, 2014, 124:14-27.
5. Spivak, A.; Garrison, W.H.; Nguyen, J.P. Review of an immiscible CO2 project, tar zone, fault block V, Wilmington field, California. SPE Reserv. Eng. 1990, 5, 155–162.
6. Todd, M.; Grand, G. Enhanced oil recovery using carbon dioxide[J]. Energy Conv. Manag. 1993, 34, 1157–1164.
7. Farajzadeh R . Enhanced transport phenomena in CO2 sequestration and CO2 EOR:[J]. Civil Engineering & Geosciences, 2009.
8. Tovar F D , Barrufet M A , Schechter D S . [Society of Petroleum Engineers SPE Improved Oil Recovery Conference - Tulsa, Oklahoma, USA (2018-04-14)] SPE Improved Oil Recovery Conference - Gas Injection for EOR in Organic Rich Shale. Part I: Operational Philosophy[J].
9. Iglauer S , Sarmadivaleh M , Al-Yaseri A , et al. Permeability evolution in sandstone due to injection of CO2-saturated brine or supercritical CO2 at reservoir conditions[J]. Energy Procedia, 2014, 63:3051-3059.
10. Al-Yaseri A Z , Zhang Y , Ghasemiziarani M , et al. Permeability Evolution in Sandstone due to CO2 Injection[J]. Energy & Fuels, 2017:acs.energyfuels.7b01701.
11. Yellig, W.; Metcalfe, R. Determination and Prediction of CO2 Minimum Miscibility Pressures. J. Pet. Technol.1980, 32, 160–168.
12. Tran, T.Q.M.D.; Neogi, P.; Bai, B. Stability of CO2 Displacement of an Immiscible Heavy Oil in a Reservoir. SPE J. 2017, 22.
13. Jinsheng Zhao, Xiao An, Shouwu Qiao. The extraction effect of CO2 injection on the flow properties of crude oil[J]. Petroleum Science and Technology, 2019,37(6), 710-717.

Experimental and Numerical Studies on Asphaltene Precipitation Distribution during CO2 Flooding in Ultra-Low Permeability Reservoirs

Authors:
Mu, L., China University of Petroleum - Beijing
Zou, J., China University of Petroleum - Beijing

Asphaltene precipitation (AP) is an aspect of serious causes in decrease of production capacity of mature oil fields and is a common phenomenon during CO_2injection in ultra-low permeability reservoirs. Therefore, there is an urgent need to determine precipitation mechanism, related influencing factors, and possible distribution area.

Laboratory experiments and numerical simulations are conducted to study the influencing factors and distribution mechanism of AP. A visible detection system was employed to obtain the critical pressure and temperature of AP in the CO_2-oil system. Core flooding experiments were designed to study the influencing factors of AP under different pressure and temperature on porosity, permeability, and oil recovery efficiency. Prior to constructing the three-dimensional numerical model, a one-dimensional model is built to fit the core flooding experiments, thereby ensuring accuracy.

The results obtained by experimental studies indicate that the amount of AP increases as the pressure rises to minimum miscibility pressure and then decreases slowly. As the temperature increases, the amount of deposition also increases and then remains almost unchanged. Additionally, the numerical simulation indicates that AP is affected by three main factors: oil components, pressure, and temperature conditions. In turn, AP affects porosity, permeability, and oil recovery.

The encouraging results of the field application effect of deposition distribution prediction based on the numerical model corrected by displacement experiments can enable efficient and referable solutions regarding reduction of AP in CO_2flooding and improvement in ultimate oil recovery.

Experimental Investigation of Supercritical CO2 Injection for Enhanced Gas Recovery in Tight Gas Reservoir

Authors:
Ding, J., Chengdu University of Technology
Cao, T., SINOPEC North China Company
Wu, J., SINOPEC North China Company

Tight gas reservoirs always have low natural gas recovery due to poor physical properties. However, there is still no effective means to enhance gas recovery (EGR) in tight gas reservoir. In this paper, experimental approaches were carried out to study the supercritical CO_2 injection for EGR in tight gas reservoirs. Phase behavior investigation was performed to indicate the property difference between supercritical CO_2 and CH_4 under reservoir condition. Results show that supercritical CO_2 has significantly higher density and viscosity than CH_4 under reservoir condition. Gravity differentiation and near-piston displacement can be achieved in case of supercritical CO_2 injection and thus the displacement efficiency can be improved. Gas adsorption tests show that the adsorption capacity of supercritical CO_2 in tight sandstone is more than 60% higher than that of CH4. The lower the permeability is, the stronger the adsorption capacity of supercritical CO_2 become, which means supercritical CO_2 can replace CH_4 in tight cores and EGR through competitive adsorption. Diffusion simulation show that supercritical CO_2 shows weak diffusion capacity and slow diffusion process in CH_4, which indicates that supercritical CO_2 and CH_4 are not easy to mix and it is good for displacement and EGR. On the basis of these fundamental tests, long-core experiments of supercritical CO_2 injection for EGR in tight gas reservoir were carried out using natural cores from DS tight gas reservoir. Results indicate that supercritical CO_2 injection can improve gas recovery by 13.5% on the basis of gas depletion. Furthermore, the influencing factors of supercritical CO_2 injection in tight cores were experimentally investigated. Results demonstrate that supercritical CO_2 injection have better EGR effect in case of lower permeability, higher water saturation and greater dip angle.

Experimental Modeling of Gas Channeling for Water-alternating-gas flooding in High-temperature and High-pressure Reservoirs

Yunbo Li[1], Yang Yao[2], Mei Qi[1], Chuang Zhao[2], Bin Han[1], Zhaojie Song[2,], Likun Xu[1], Qiang Liu[2], Jirui Hou[2], Liguang Song[2,†]*

1. Research Institute of Petroleum Exploration & Development, CNPC, China;

2. Unconventional Petroleum Research Institute, China University of Petroleum, Beijing

Abstract

The objective of this study was to investigate gas channeling behavior and its governing factors during water-alternating-gas (WAG) flooding in high-temperature and high-pressure reservoirs. According to the geological characteristics and development conditions of the target carbonate reservoir, various core models were fabricated and the relevant experimental method was designed based on the similarity criterion of physical simulation. A series of experiments were performed to discussed the production performance of WAG flooding, and the effect of core heterogeneity and injection rate on oil displacement efficiency was investigated. Experimental results showed that WAG flooding presented great potential of enhanced oil recovery and good effect on delaying water and gas production. The severer the core heterogeneity, the smaller the fluctuation range and the less the fluctuation frequency of water cut and gas-oil ratio. The effect of WAG flooding on water and gas production control became worse with an increase in core heterogeneity. The injection rate presented less significant effect on oil recovery compared to core heterogeneity. Water and gas breakthrough occurred earlier at a higher injection rate. This experimental study could provide guiding suggestions for reasonable implementation of WAG flooding in the oilfield.

Keywords: WAG flooding; gas channeling; enhanced oil recovery; high-temperature and-high pressure reservoir; physical simulation

1. Introduction

As a non-renewable energy, petroleum plays an increasingly important role in industry[1]. As most of the oil fields have entered the later stage of production, the water cut has increased greatly and the oil production has decreased sharply. Unfortunately, new fields are becoming harder to explore and harder to exploit. In addition, the development and utilization of new alternative energy are in the exploration stage. Therefore, enhanced Oil Recovery (EOR) is still a leading technology for the high-efficiency exploitation of the

* Corresponding author, songz@cup.edu.cn (Z. Song).
† Presenting author.

oilfield[2]. Among many EOR technologies, CO_2 flooding has become increasingly popular due to the unique advantages of CO_2 and crude oil with the low minimum miscible pressure (MMP) and its ability of being soluble in crude oil and reducing viscosity. In addition, CO_2 is a greenhouse gas. Injecting it into reservoirs can not only realize long-term underground storage of CO_2 and reduce the greenhouse effect, but also improve oil recovery. However, in the process of oil-displacement, there are also many problems[3-6]. For example, in the middle and late stage of displacement, gas channeling may occur. When gas channeling occurs, gas production rises sharply, and the amount of oil produced is very small. Hence, in recent years WAG flooding has been studied in laboratories and oilfields because of its ability to delay water and gas channeling. Heterogeneity is a universal characteristic of reservoirs. Reservoir heterogeneity is usually characterized by permeability ratio[7]. The existing high permeability layers, artificial or natural fractures, can make the formation more heterogeneous to some extent. Heterogeneity is a sensitive factor affecting oil recovery. In this study, a series of experiments were performed to discussed the production performance of WAG flooding, and the effect of core heterogeneity and injection rate on oil displacement efficiency was investigated.

2. Experimental section

2.1 Preparation of core samples

The cores used in this research were artificial carbonate core models with a sized of 4.5x4.5X30 cm, which have been widely applied in laboratory displacement experiments. Three layered heterogeneous cores with different heterogeneities were employed. As show in **Fig.1**, each core had two layers, the low permeability layer (LPL) and high permeability layer (HPL) [8-9]. Permeability contrast (PC), defined as the permeability ratio of the HPL and LPL, was chosen as the index to represent the heterogeneity of cores. The core permeability ratio V_K is 5, 10 and 15 respectively. In addition, four cores with a permeability difference of 5 are used to study the effect of injection rate on oil recovery.

Fig.1 Core samples (Size=4.5cm×4 .5cm×30 cm.)

2.2 Experimental fluid

The water used in the experiments was prepared according to the salinity of the water in the target reservoir. The salinity of simulated water is 265,662 mg/l. The oil used in the experiments was a mixing of the dead oil obtained from target oilfield and gas (CO_2:N_2=10:1) to simulate its experimental viscosity and density according to the criterion of pressure similarity and fluid parameter equivalent. The oil viscosity was 1.079 mPa·s under 84℃

temperature condition and the density was 0.847 g/cm3. Meanwhile, Injected gas is composed of CO_2 and N_2 and the ratio of both is 10:1. The compositional analysis of the dead oil by gas chromatography is presented in **Table 1**.

Table 1. Component of dead oil

Component	Wt%	Component	Wt%
C_3H_8	0.41	FC_{11}	4.71
IC_4	0.24	FC_{12}	4.26
NC_4	0.86	FC_{13}	4.44
IC_5	0.7	FC_{14}	3.61
NC_5	1.33	FC_{15}	3.39
FC_6	3.1	FC_{16}	2.72
FC_7	6.17	FC_{17}	2.28
FC_8	8.06	FC_{18}	2.47
FC_9	6.66	FC_{19}	2.2
FC_{10}	5.45	C_{20+}	36.91

2.3 Experimental setup and procedures

2.3.1 Experimental setup

Fig. 2 shows the details of the experimental setup used for WAG flooding in this study.

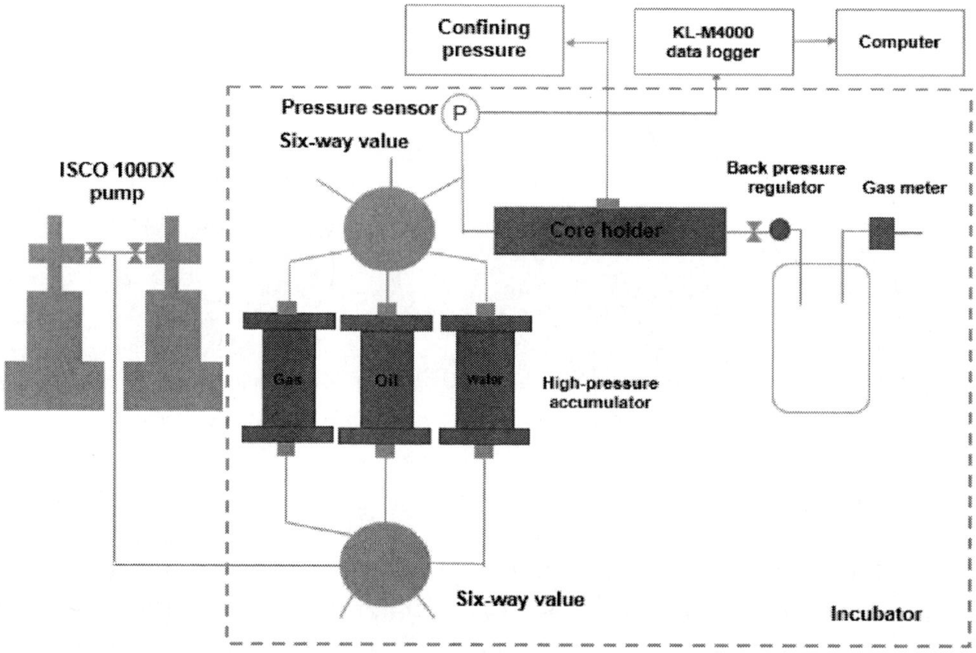

Fig. 2. Diagram of WAG flooding experimental setup

The flow chart of the experimental setup was composed of a power driving system, a core holder and a data acquisition system. An ISCO 100DX cylinder pump provided non-stop power for injecting gas/water stored in high-pressure accumulators. The core holder, high-pressure accumulator and back-pressure regulator were placed inside an oven at a temperature of 84℃. Core of saturated oil under confined water conditions was located in the core holder covered by a rubber sleeve. A confining pressure was applied to the core gripper through the rubber sleeve, which enabled the core to be strongly coated and enabled fluid to pass through the cross-sectional area of the core in the horizontal direction, preventing fluid flow around the core holder. The back-pressure regulator guaranteed the stability of the test pressure, and compressed water was used to maintain the back pressure. All of the pressures were tested and recorded by pressure sensors connected to the facility. During displacement, the tube is used to measure the volume of oil and water produced, and the volume of gas produced is measured by a gas meter.

Using the above facilities, WAG tests were performed under varying experimental conditions and injection strategies to investigate the following parameters: 1) permeability ratio (Vk=5,10,15); 2) injected rate (v=0.1ml/min, 0.2ml/min, 0.3ml/min, 0.4ml/min). The influences of such factors on WAG injection EOR processes in target reservoirs were experimentally researched.

2.3.2 Experimental procedures

An artificial carbonate core whose dimensions, permeability ratio, injected rate and back-pressure are given in **Table 2** was wrapped with a mixture of epoxy resin and curing agent.

Table 2. Core properties and experimental conditions of WAG flooding.

Core number	Physical model size	Permeability ratio	Injected rate (ml/min)	Back-pressure (MPa)
#1	4.5x4.5x30cm	5	0.2	25
#2		10	0.2	25
#3		15	0.2	25
$1		5	0.1	25
$2		5	0.3	25
$3		5	0.4	25

For core flooding experiments, the sequence of experiments was as follows.

①Artificial carbonate cores coated with epoxy resin and curing agent were put into the core holder and then confining pressure was added. ②The core was vacuumed for 4-6 hours and saturated with formation water to calculate the permeability of water phase. ③Calculate oil saturation after the core was saturated with crude oil. ④The aging stage of the oil, which lasted 24 hours, was designed to be consistent with the form of crude oil that has been stored in the actual formation for countless years. ⑤According to experimental requirements, different permeability ratio and injection rates were set. ⑥In the way of first gas injection and then water injection cycle alternating injection, the gas-water volume ratio was 1:1. ⑦The experiment was terminated when the producing gas-oil ratio reached 3000mL/mL.

3. Results and discussion

3.1 Production performance during WAG flooding

Fig. 3 shows the production performance of core #1 during WAG displacement, which can be divided into four stages: no liquid stage, evolved gas and waterless production stage, oil-gas-water co-production stage and gas channeling stage. For the other cores, the experiments under different injected rate or rock heterogeneity conditions, exhibited the similar characteristics with the #1. As illustrated in **Fig.3**, at no liquid stage, the injected fluid mainly plays the role of replenishing formation energy and reducing the viscosity of crude oil. In addition, there is mass transfer between gas and crude oil during the process to produce near-miscible or miscible effects. The crude oil is produced mainly at evolved gas and waterless production stage and oil-gas-water co-production stage. However, the oil recovery basically does not increase when gas channeling occurs.

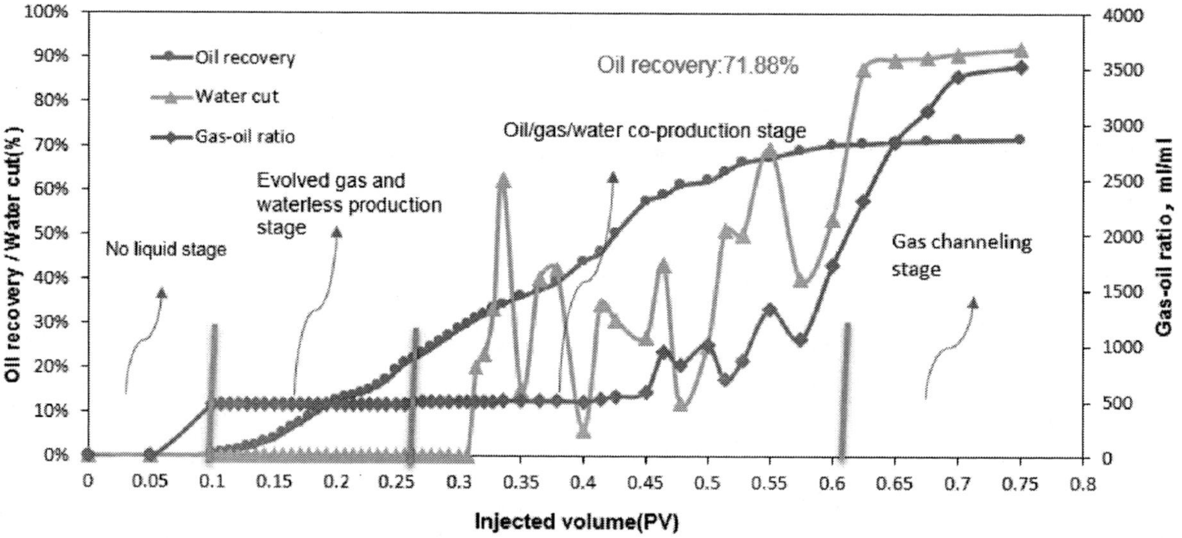

Fig.3 Producing gas-oil ratio, water cut and oil recovery curve of WAG in heterogeneous core sample (#1)

3.2 Effect of permeability heterogeneity

The effects of different permeability, temperature and pressure on the channeling of injected fluid in porous media. Through investigation, the following conclusions are drawn: for cores with a great difference in permeability between the low permeability layer and the high permeability layer, injection fluid tends to be in the high permeability layer. In other words, the high permeability layer is easier to form channeling channels than the low permeability layer. In addition, the higher the permeability of the reservoir is, the more obvious the phenomenon of gas overlap will be, and the more serious the gas/water channeling will be.

3.2.1 Effect of heterogeneity on oil recovery

WAG injection tests were performed in cores with different heterogeneities to determine the influence of heterogeneity on oil recovery. Permeability ratio was chosen as the index of heterogeneity. It was designed to be 5, 10, and 15 according to the reservoir heterogeneities of the target oilfield. A permeability ratio of 5 represents a weak heterogeneity, while 10 permeability ratio represents a moderate heterogeneity, and 15 permeability ratio simulates strong heterogeneity. As explained in the experimental section, WAG displacement experiment was carried out under the same conditions except for different permeability ratio, and the experimental results were shown in **Fig. 4-6**.

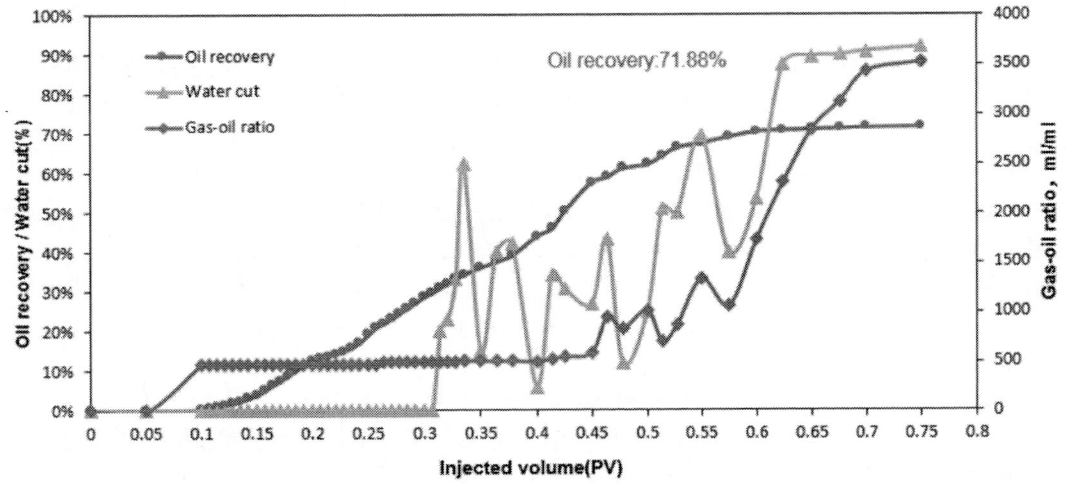

Fig.4 Producing gas-oil ratio, water cut and oil recovery curve of WAG flooding in heterogeneous core sample (#1)

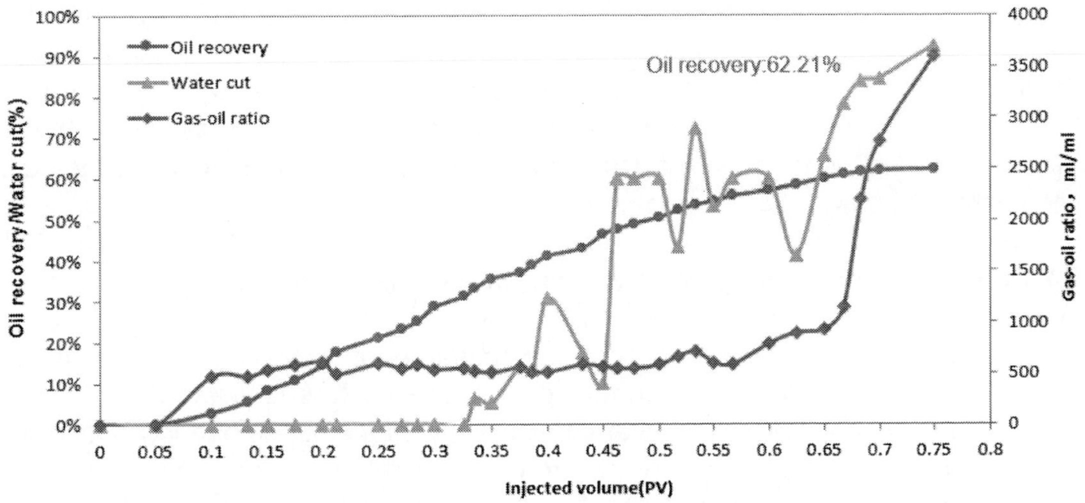

Fig.5 Producing gas-oil ratio, water cut and oil recovery curve of WAG flooding in heterogeneous core sample (#2)

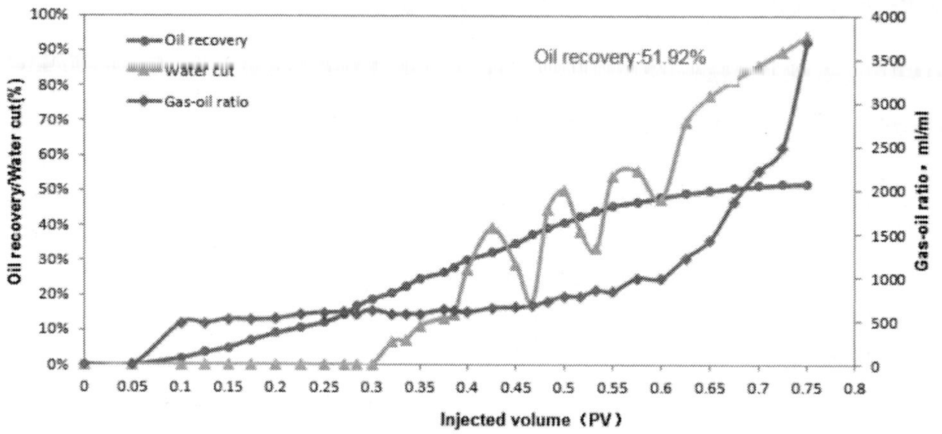

Fig.6 Producing gas-oil ratio, water cut and oil recovery curve of WAG flooding in heterogeneous core sample (#3)

As shown in the **Fig.4-6**, other conditions being unchanged, oil recovery with a permeability ratio of 5 was 71.80%, a permeability ratio of 10 was 62.21%, and a permeability ratio of 15 was 51.92%. In terms of oil recovery, the ultimate recovery decreased with the rise of permeability ratio. In other words, with the increase of core heterogeneity, the displacement effect of WAG became worse. This indicates that the injected fluid flowed preferentially into the highly permeable layer and bypassed the low permeable layer, resulting in reduced sweep efficiency and affecting ultimate recovery.

As far as water cut and gas-oil ratio were concerned, no matter the core is weak heterogeneity, moderate heterogeneity, or even strong heterogeneity, water cut and gas-oil ratio show roughly the same trend with the increase of injection volume, and they all rise in a wavy zigzag manner during WAG displacement. However, the higher the heterogeneity, the smaller the fluctuation range of water cut and gas-oil ratio, and the less the fluctuation frequency. That can be explained that the effect of WAG displacement on controlling water and adding oil became worse as core heterogeneity increases. In the later stage of displacement, the water content and gas-oil ratio significantly increased and no longer fluctuated, indicating the formation of gas channeling, the recovery curve tended to be stable and the alternating gas-water flooding gradually failed.

3.2.2 Effect of heterogeneity on the timing of gas breakthrough

During alternating gas/water injection, the increase of water injection pressure was obvious, while the increase rate of gas injection pressure was low, and the displacement pressure difference presented the characteristics of stage supercharging. However, the injection pressure also showed an upward trend after several cycles. It was caused by the large amount of air/water interface, the rapid increase of capillary resistance, and the gradual rise of gas injection pressure. Therefore, WAG injection can be used as an effective method to control gas channeling. However, with the increase of permeability ratio, the effect of WAG flooding to control gas channeling began to decline. **Table 3** showed the gas breakthrough time, gas channeling time and their oil recovery under different permeability ratio.

Table 3. Oil recovery at different displacement times

Permeability ratio	Gas breakthrough timing (PV)	Oil recovery at gas breakthrough timing	Gas channeling timing (PV)	Oil recovery at gas channeling timing
5	0.264PV	22.19%	0.6PV	70.63%
10	0.175PV	11.04%	0.667PV	60.91%
15	0.225PV	10.68%	0.65PV	50%

When the permeability ratio became larger and larger, that is, core heterogeneity became stronger and stronger, the gas breakthrough time tended to advance. Due to the limitation of the number of experimental groups, there was no obvious regularity. However, according to the parameter of oil recovery at gas breakthrough, the oil recovery went from 22.19% to 10.68%, This phenomenon is relatively intuitive, that is, with the enhancement of heterogeneity, the oil recovery at the gas breakthrough timing gradually dropped off. As can be seen from **Table 3**, cores with permeability ratio of 5 and 10 had 10.15% difference in recovery at gas breakthrough, while cores with permeability ratio of 10 and 15 had only 0.36% difference in recovery. It can be inferred that the sensitive limit of the recovery at the gas breakthrough moment to the permeability ratio may exist, of course, it needs to be further studied. In addition, no obvious regularity was found at the time of gas channeling.

3.2.3 Effect of heterogeneity on water cut and gas-oil ratio

Fig.7 illustrated the production performance of water cut with oil recovery in different ratio. With the increase of oil recovery, the water cut also presented an upward trend, and when the recovery reached a certain level, the water cut rised sharply. The water cut curve showed a trend of shifting to the right and became more concave with the increase of the permeability ratio. To a certain extent, it can also reflect that the water control effect of WAG flooding gradually weakened with the enhancement of heterogeneity.

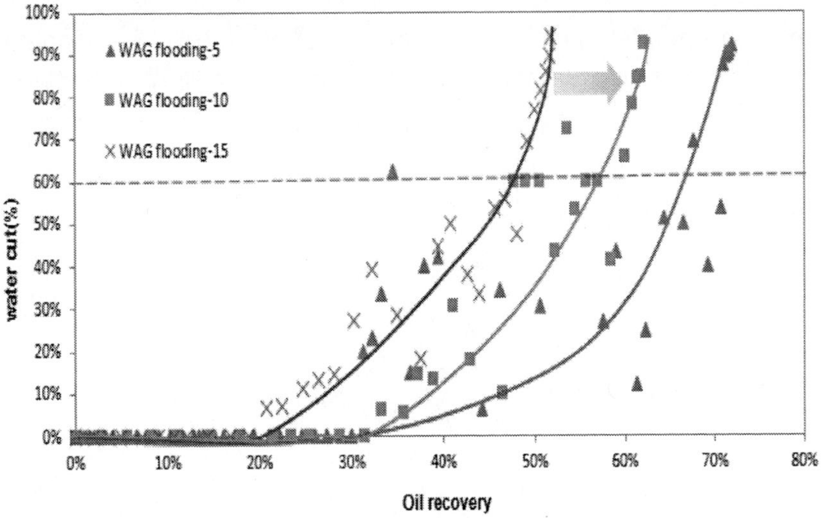

Fig.7 Production characteristics of water cut with oil recovery

Fig. 8 depicted the production performance of gas-oil ratio with oil recovery in diverse heterogeneity. The rising curve of gas-oil ratio was generally consistent with that of water cut in **Fig.7.** With the increase of heterogeneity, the upward curve of gas-oil ratio shifted to the right and the ability to delay gas channeling decreased gradually.

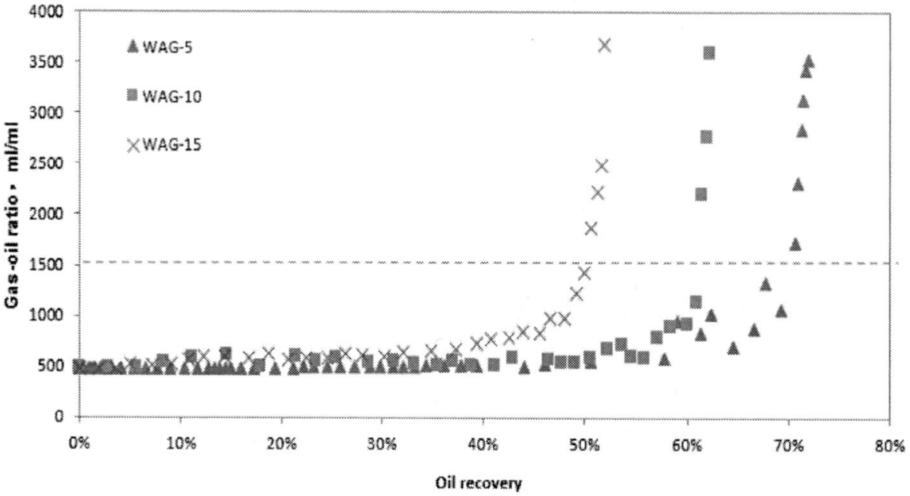

Fig.8 Production characteristics of gas-oil ratio with oil recovery

According to the comprehensive analysis of the results in Fig. 7 and Fig. 8, if the medium water cut was 60%, the recovery of weak heterogeneous, moderate heterogeneous and strong heterogeneous cores was 67.63%, 58.41% and 48.18% respectively. Using the same method, with the gas-oil ratio of 1500 as the boundary, the oil recovery was respectively 70.63%, 60.91% and 50% with the core heterogeneity from weak to strong. From the change of water cut and gas-oil ratio, it can be inferred that the effect of controlling water and delaying gas channeling of WAG flooding decreased with the increase of heterogeneity.

3.3 Effect of injection rate

The displacement effects of four different injection rates in heterogeneous cores were studied. The production characteristics of oil, gas and water were shown in **Fig. 9-12**.

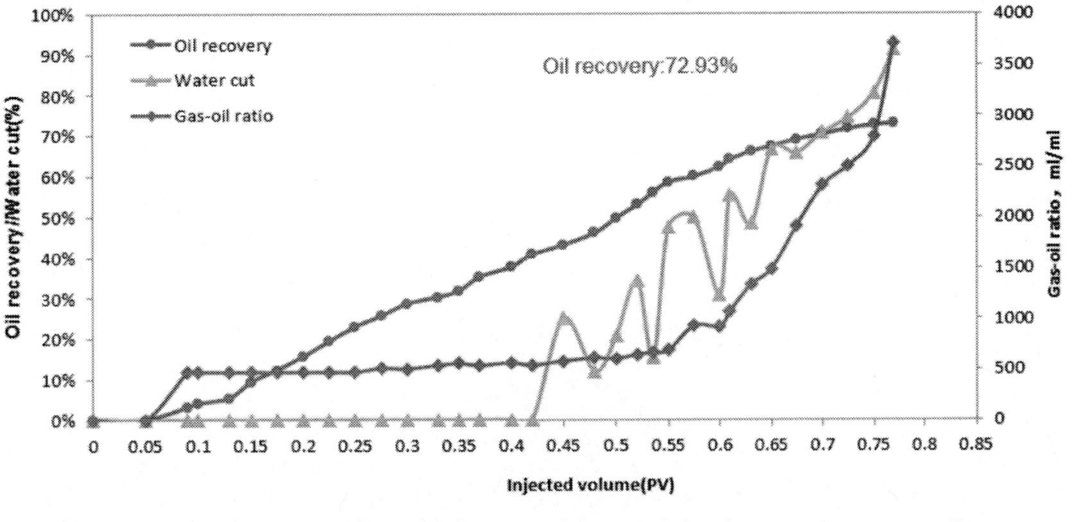

Fig.9 Producing gas-oil ratio, water cut and oil recovery curve of WAG flooding in heterogeneous core sample ($1)

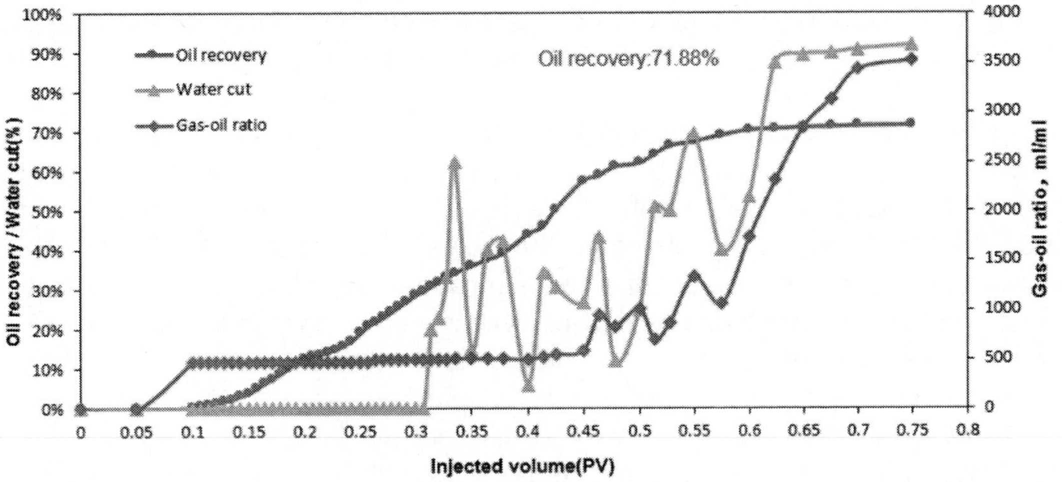

Fig.10 Producing gas-oil ratio, water cut and oil recovery curve of WAG flooding in heterogeneous core sample (#1)

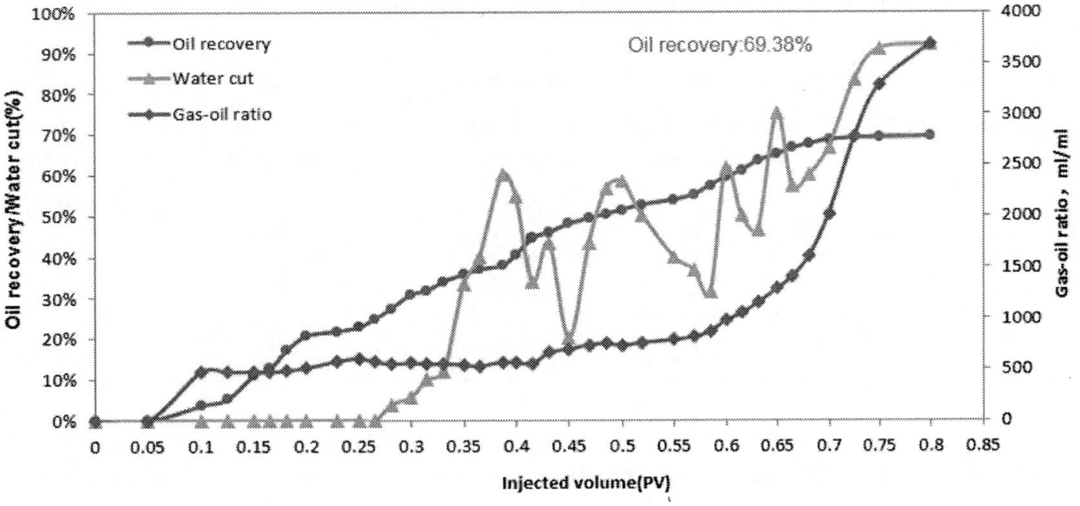

Fig.11 Producing gas-oil ratio, water cut and oil recovery curve of WAG flooding in heterogeneous core sample ($2)

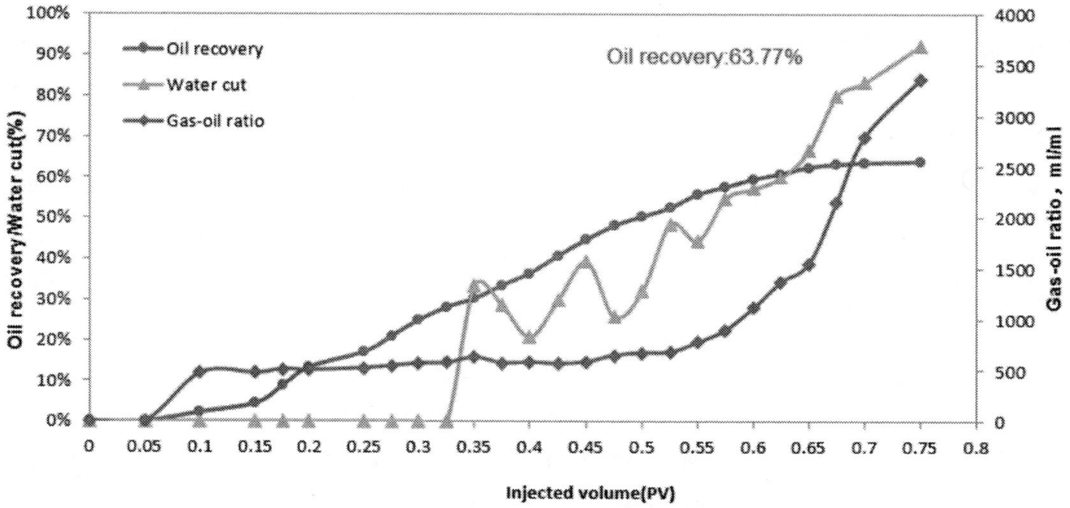

Fig.12 Producing gas-oil ratio, water cut and oil recovery curve of WAG flooding in heterogeneous core sample ($3)

In the range of experimental injection rate, with the increase of displacement rate, the oil recovery decreased gradually. As can be seen from the Fig.9-12, when the injected rate is 0.1ml/min, the oil recovery is the highest, reaching 72,93%. However, the displacement time is twice that of 0.2ml/min. According to the actual situation on the site and the overall economic benefits, the experiment of 0.1ml/min cannot reach the maximum NPV. In the gas production stage, the growth rate of recovery was the slowest when the injection rate was 0.4 ml/min, and the fluctuation range of gas-oil ratio and water cut was the most obvious when the injection rate was 0.2 ml/min, and the increase of recovery was the largest at this time.

Table 4 depicted the oil recovery at different displacement times. The increase of injection speed led to the advance of gas breakthrough time, and the recovery efficiency decreased at both gas breakthrough time and gas channeling time.

Table 4 Oil recovery at different displacement times

Injected rate	Gas breakthrough timing (PV)	Oil recovery at gas breakthrough timing	Gas channeling timing (PV)	Oil recovery at gas channeling timing
0.1 ml/min	0.3PV	28.53%	0.7PV	70.27%
0.2 ml/min	0.264PV	22.19%	0.6PV	70.63%
0.3 ml/min	0.181PV	17.11%	0.665PV	66.49%
0.4 ml/min	0.175PV	8.83%	0.65PV	62.47%

The relationship between water cut and oil recovery was shown in **Fig.13**. The water cut curve tended to shift to the right with the increase of injection rate, but for a single water cut curve, the distribution of data points was complex, and no obvious rule was found.

Compared with **Fig.13**, the gas-oil ratio changed more obviously with the increase of oil recovery in **Fig.14**. As can be seen from the Fig.14, the gas-oil ratio with injection rate of 0.2ml/min rised the slowest.

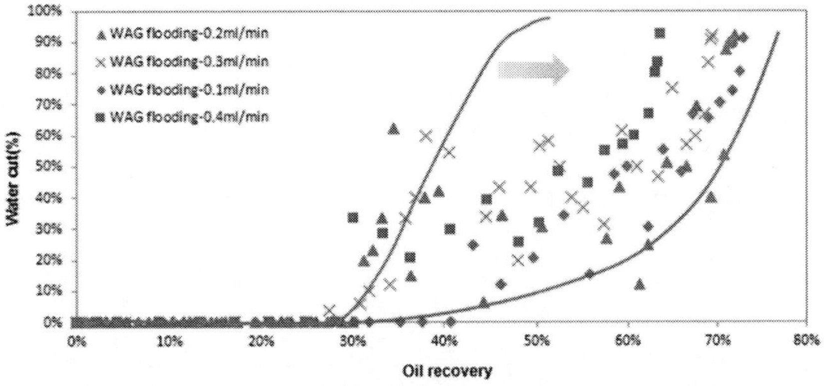

Fig.13 Production characteristics of water cut with oil recovery

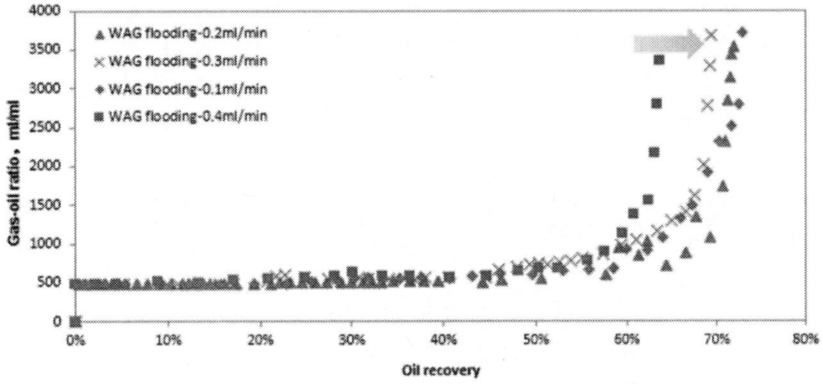

Fig.14 Production characteristics of gas-oil ratio with oil recovery

Compared with the permeability ratio, the injection rate had little effect on the ultimate oil recovery, mainly affecting the WAG flooding effect time. The faster the injection rate was, the earlier the effect time would be. Conversely, the slower the injection rate, the later the effect time. For different reservoirs, there was an optimal injection rate. If the injection rate was too fast, the formation of gas or water channeling would be intensified, and the displacement effect of WAG would be worse.

4. Conclusions

1) WAG flooding achieved an oil recovery of 71.88%, 62.21% and 51.92% in the core samples with a permeability ratio of 5, 10 and 15, respectively. With the increase of core heterogeneity, oil recovery decreased remarkably.

2) The severer the core heterogeneity, the smaller the fluctuation range and the less the fluctuation frequency of water cut and gas-oil ratio. The effect of WAG flooding on water and gas production control became worse with an increase in core heterogeneity. In the later

stage of displacement, the water cut and gas-oil ratio significantly increased and no longer fluctuated, and the recovery tended to be stable.

3) The injection rate presented less significant effect on oil recovery compared to core heterogeneity. Water and gas breakthrough occurred earlier at a higher injection rate.

Acknowledgments

This work is sponsored by the project of RIPED-2018-JS-688.

References

[1]. Wang, L., et al., Advances in improved/enhanced oil recovery technologies for tight and shale reservoirs. Fuel, 2017. 210: 425-445.

[2]. Song, Z., et al., Conformance control for CO_2-EOR in naturally fractured low permeability oil reservoirs. Journal of Petroleum Science and Engineering, 2018. 166: 225-234.

[3]. Alfarge, D., M. Wei and B. Bai, Data analysis for CO_2-EOR in shale-oil reservoirs based on a laboratory database. Journal of Petroleum Science and Engineering, 2018. 162: 697-711.

[4]. Ren, B., et al., Monitoring on CO_2 migration in a tight oil reservoir during CCS-EOR in Jilin Oilfield China. Energy, 2016. 98: 108-121.

[5]. Lashgari, H.R., et al., Evaluation of carbon dioxide storage and miscible gas EOR in shale oil reservoirs. Fuel, 2019. 241: 1223-1235.

[6]. Ma, Q., et al., Experimental investigation on the influence factors and oil production distribution in different pore sizes during CO_2 huff-n-puff in an ultra-high-pressure tight oil reservoir. Journal of Petroleum Science and Engineering, 2019. 178: 1155-1163.

[7]. Yang, J., et al., Experimental studies on CO_2 foam performance in the tight cores. Journal of Petroleum Science and Engineering, 2019. 175: 1136-1149.

[8]. Bikkina, P., et al., Influence of wettability and permeability heterogeneity on miscible CO_2 flooding efficiency. Fuel, 2016. 166: 219-226.

[9]. Ding, M., et al., Oil recovery from a CO_2 injection in heterogeneous reservoirs: The influence of permeability heterogeneity, CO_2 -oil miscibility and injection pattern. Journal of Natural Gas Science and Engineering, 2017. 44: 140-149.

FEASIBILITY STUDY OF GAS-EOR USING CO_2 AND N_2 MIXTURE IN A HEAVY OIL RESERVOIR: EXPERIMENTS AND PILOT TEST

Dr. Hongda Hao, China University of Petroleum (Beijing), China
Dr. Jirui Hou, China University of Petroleum (Beijing), China
Dr. Fenglan Zhao, China University of Petroleum (Beijing), China
Dr. Handong Huang, China University of Petroleum (Beijing), China
Mr.Zhixing Wang, China University of Petroleum (Beijing), China
Dr. Huaizhu Liu, Jidong Oilfield Company, CNPC, China

Introduction

Since shallow-buried oil reservoirs of Nanpu Block, Jidong Oil Field, China belong to complex fault-block oil reservoirs, effective flooding networks are difficult to be established compared with other conventional oil reservoirs. Although horizontal wells are widely used in the field to expand drainage area, severe water invasions still happened after years of development **[1,2]**. Cyclic CO_2 injection, also known as CO_2 huff-n-puff, was then conducted to enhance oil recovery (EOR) since 2010, and great profits has been brought with 109.5×10^4 t of oil productions until the year of 2018. As a kind of solvent, CO_2 can dissolve with the heavy oil, and then cause oil swelling and viscosity reduction **[3, 4, 5, 6, 7]**. However, production problems such as wellbore corrosions, gas channeling, etc. still cannot be avoided during the operation of CO_2 injection, which then reduce the oil yielding and lower the CO_2 utilization. N_2 can also be used to enhance oil recovery according to the literatures **[8, 9, 10]**, more importantly, it is not corrosive and usually cheaper than CO_2 for EOR applications. Thus, CO_2 and N_2 mixture is proposed as an alternative gas-EOR technique for those shallow-buried heavy oil reservoirs.

In order to figure out the interaction mechanisms between CO_2/N_2 mixture and heavy oil, a feasibility study is firstly evaluated using a PVT cell in the laboratory. Mixtures with different CO_2/N_2 molar ratios (1:0, 4:1, 7:3, 1:1, 0:1) are designed, and the high-pressure properties of gas/oil system including saturation pressure, volume factor and viscosity are investigated. Then, a series of cyclic gas injection experiments are conducted using outcrop cores to compare the oil recovery factors of CO_2, N_2, and CO_2/N_2 mixtures. A pilot test of CO_2/N_2 mixture injection is also introduced in this paper.

Experiments

PVT analysis for gas/oil system

A PVT cell and a viscometer are used in the laboratory as shown in **Fig.1** and **Fig.2**. The operation pressure and temperature of the PVT cell is 200 MPa and 200 °C, and the viscometer's measurement range is from 0.3 mPa·s to 20000 mPa·s. The oil and water samples are collected from one block of Nanpu Oil Field, China. The density of formation oil is 0.89 g/cm^3, and the oil viscosity is 52.13 mPa·s under formation conditions (65 °C, 16.24 MPa). Both of the injected CO_2 and N_2 are with purities of 99.99 mol%.

The CO_2/N_2 molar ratio is set as 1:0, 4:1, 7:3, 1:1, 0:1, respectively, and the procedures for PVT analysis are as follows. (1) The PVT cell and the viscometer are cleaned and evacuated using a vacuum pump. (2) A specific volume of formation oil is injected into the cell and the viscometer, followed by a specific volume of gas. (3) Increase the pressure and temperature of the cell and the viscometer, then stir the gas/oil mixture for 12 h. The viscosity of formation oil (μ_o) can be measured at formation conditions

(65 °C, 16.24 MPa). (4) Decrease the pressure step by step, and record the oil volume at a specific pressure. The saturation pressure (P_b) of the gas/oil system can be determined when the first gas bubble is observed in the PVT cell. (5) Change the injecting gas/oil ratio to another value, and repeat (1) to (4), then another P_b value and μ_o value can be obtained. (6) After the changes of gas/oil ratio for five to eight times, a series of P_b values and a series of μ_o value can be obtained as the changes of CO_2/N_2 ratio. (7) During the experiment procedure, the oil and gas volumes under different pressures can also be recorded, and then a series of volume factor (B_o) values versus different gas/oil ratios can be obtained.

Fig.1. Picture of PVT analysis apparatus

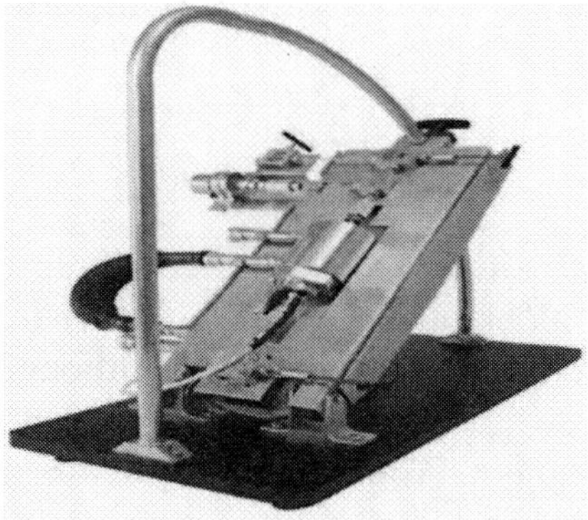

Fig.2. Picture of the viscometer

Cyclic gas injection experiments

cyclic gas injection experiments are designed to compare the oil recovery factors of CO_2, N_2 and CO_2/N_2 mixture, and the experimental setup is shown in **Fig.3**. The cores used in the experiments are outcrop cores with an average size of 300×45×45 mm^3, the average permeability of the core is 497.3×10^{-3} μm^2, and the average porosity is 17.02%. The formation oil and water samples are collected from the block, and the injected CO_2 and N_2 are with purities of 99.99 mol%.

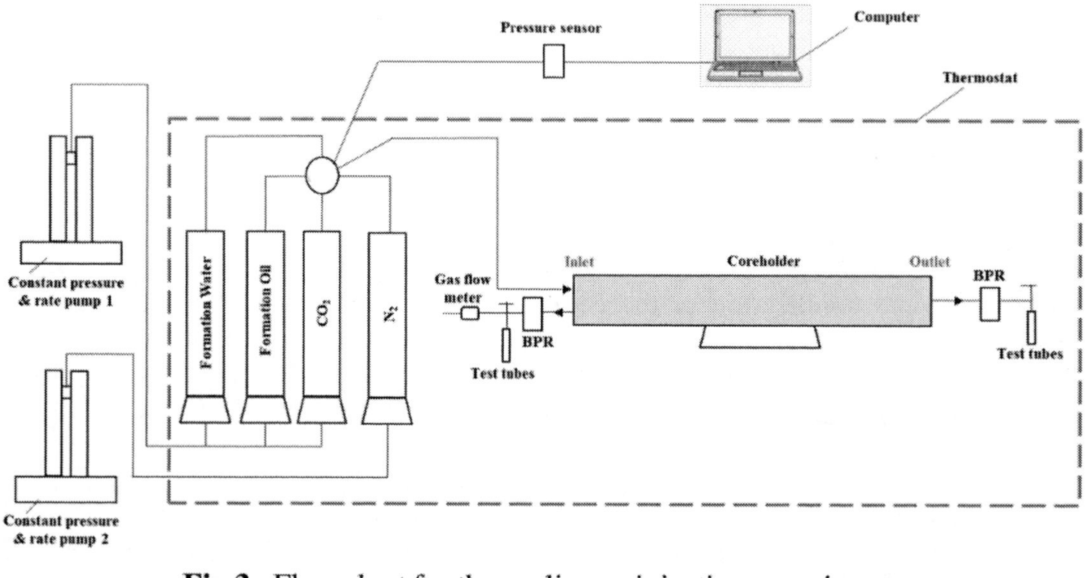

Fig.3. Flow chart for the cyclic gas injection experiments

Some preparations are conducted before the experiments as follows. (1) Epoxy resins are coated on the surface of the core to avoid CO_2 corruption, then the bulk volume of the core is measured. (2) The core is evacuated using a vacuum pump. After the core is saturated with formation water, the porosity is determined as the ratio of brine saturation volume to the bulk volume. (3) Brine is injected into the core to measure the permeability. (4) The core is displaced by oil to reach a residual water saturation, and the initial oil saturation is calculated as the ratio of injected oil volume to the pore volume.

CO_2, N_2 and CO_2/N_2 mixtures are injected into the core separately, and the procedures for the gas injection experiments are as follows. (1) The initial temperature is set as 65 °C, and the initial pressure is set as 5 MPa using a back pressure regular (BPR). (2) Gas is injected into the core with a rate of 0.3 mL/min until the injection volume reaches 0.05 PV (Under formation conditions). (3) The inlet is shutoff with a soaking time of 12 h, and then opened to start a production process. When the pressure drops to 5 MPa again, one cycle is terminated. (4) Repeat procedure (2) and procedure (3) for another three times, and the whole experimental process is terminated after four cycles of gas injection. The pressure, the production of oil and gas are measured during the experiments. (5) Change the injecting CO_2/N_2 ratio as 1:0, 4:1, 7:3, 1:1 and 0:1, and repeat the procedures above, then the oil recovery factor for different gas injecting media can be calculated.

Results and discussion

PVT comparisons of CO_2/oil, N_2/oil, and CO_2/N_2/oil systems

Fig.4 shows the plot of saturation pressure (P_b) versus injecting gas molar ratio for the gas/oil system. The initial value of P_b for the formation oil is 11.2 MPa, which will raise as the increase of injecting gas ratio. P_b of the heavy oil is more sensitive with pure N_2 compared with pure CO_2. The saturation pressure influenced by CO_2/N_2 mixture is between pure CO_2 and pure N_2, which is also influenced by CO_2/N_2 molar ratio of the mixture. For example, with 20 mol% of CO_2/N_2 mixture injection, P_b influenced by the gas with a CO_2/N_2 ratio of 1:0, 4:1, 7:3, 1:1 and 0:1 is 13.6 MPa, 19.62 MPa, 22.63 MPa, 29.62 MPa and 51.39 MPa, respectively. The dissolving capacity for different injecting gas media can be also reflected through **Fig.4**. For example, when the pressure is 30 MPa, CO_2/N_2 mixture dissolved into the oil with a CO_2/N_2 ratio of 1:0, 4:1, 7:3, 1:1 and 0:1 is 12 mol%, 18 mol%, 29 mol%, 52 mol% and 57 mol%, respectively. CO_2 has a much better dissolving capacity into the heavy oil compared with N_2, which then influence the dissolving capacity of CO_2/N_2 mixture. The dissolving capacity of CO_2/N_2 mixture is enhanced dramatically as the increase of CO_2/N_2 molar ratio.

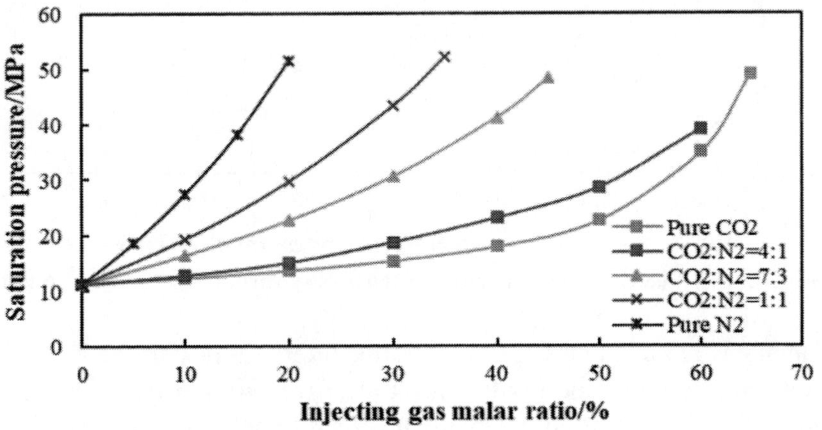

Fig.4. Saturation pressure (P_b) for the gas/oil system at 65 °C

Volume factor (B_o) is used to evaluate oil swelling capacity, which is discussed as shown in **Fig.5**. The initial value of B_o for the heavy oil is 1.058 under formation conditions, which will also raise as the increase of injecting gas malar ratio. CO_2 has a better oil swelling capacity for the heavy oil compared with N_2. The volume factor influenced by CO_2/N_2 mixture is also between pure CO_2 and pure N_2, and is closely related to CO_2/N_2 molar ratio of the mixture. B_o raises as the increase of CO_2/N_2 molar ratio, and a better oil swelling capacity can be obtained with a higher CO_2 concentration for the mixture.

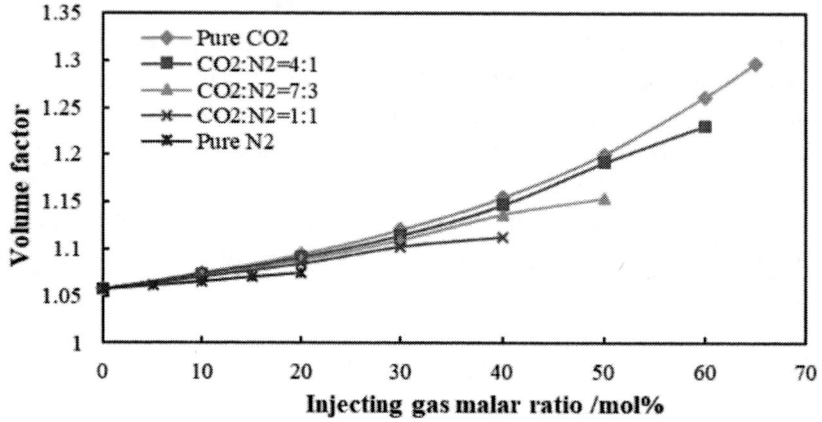

Fig.5. Volume factor (B_o) for the gas/oil system at 65 °C and 16.24 MPa

The oil viscosity (μ_o) reduction influenced by CO_2/N_2 mixture is shown in **Fig.6**. The initial value of μ_o for the formation oil is 52.13 mPa·s, which will decrease as the increase of injecting gas molar ratio. As more gas is dissolved into the formation oil, the viscosity reduction by the injecting gas is more obvious. CO_2 has a much better viscosity reduction capacity for the heavy oil compared with N_2. For the CO_2/N_2 mixture, the capacity of viscosity reduction is closely related to the CO_2/N_2 molar ratio. With a higher CO_2/N_2 ratio, a better viscosity reduction capacity will be obtained for the mixture.

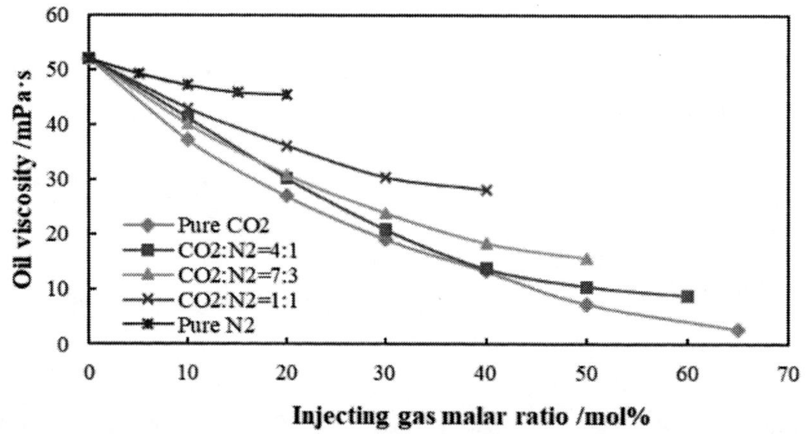

Fig.6. Oil viscosity (μ_o) for the gas/oil system at 65 °C and 16.24 MPa

The PVT analysis reveals that CO_2 has a better interaction with the heavy oil compared with N_2. For the CO_2/N_2 mixture, its interaction with formation oil is always between pure CO_2 and pure N_2. A higher CO_2/N_2 molar ratio for the mixture can always lead to better capacities of dissolution, oil swelling and viscosity reduction, which are the dominant mechanisms for the heavy oil extraction.

Recovery factor comparisons of CO_2, N_2 and CO_2/N_2 mixtures

Table 1 shows the total oil recovery factors after four cycle of gas injection using different gas media. After four cycles of gas injection, pure CO_2 and the 4:1 (CO_2/N_2 molar ratio) mixture achieve higher oil recovery factors. The oil recovery of 4:1 mixture is 17.31%, which is close to the oil recovery of 19.03% achieved by pure CO_2. 7:3 mixture achieves the middle oil recovery of 13.27%, and the 1:1 mixture and the pure N_2 achieve the lowest oil recovery factors, which are less than 10%. As discussed above, the oil extraction mechanisms for CO_2/N_2 mixture is closely related to the injecting CO_2/N_2 ratio. Since the concentration of N_2 component increases as the CO_2/N_2 ratio decreases, the interactions between gas and oil is weakened, which then affects the oil extraction of the CO_2/N_2 mixture. For the mixture with a CO_2/N_2 molar ratio of 4:1, CO_2 accounts for a large proportion of the mixture, and the superior performances of gas dissolution, oil swelling and viscosity reduction dominated by CO_2 are highly reserved for enhanced oil recovery.

Table 1. Total oil recovery of cyclic gas injection using different gas media

CO_2/N_2 molar ratio	**1:0**	**4:1**	**7:3**	**1:1**	**0:1**
Total oil recovery/%	19.03	17.31	13.27	9.09	8.09

Fig.7 shows the oil recovery for each cycle during gas injection experiments. It can be observed that although the oil recovered by 4:1 mixture is less than the oil recovered by pure CO_2 during the first and the second cycle, a better oil recovery can still be obtained for the 4:1 mixture during the third and the fourth cycle. **Fig.8** shows the pressure of cyclic gas injection using different gas media. The pressure enhancement using pure N_2 is much higher than the pressure using CO_2/N_2 mixture and pure CO_2, however, the oil recovery factor is still the poorest. This reveals that the critical factors for gas-EOR is to use enough CO_2 to interact with the heavy oil, while the pressure enhancement using N_2 could act as a supplementary role for the oil extraction. Although pure CO_2 has the best interactions with the heavy oil, 4:1 mixture can reach a higher pressure enhancement compared with pure CO_2. The gas/oil interactions dominated by CO_2 coupled with the pressure supplement by N_2 lead to a favorable oil recovery for the 4:1 mixture injection.

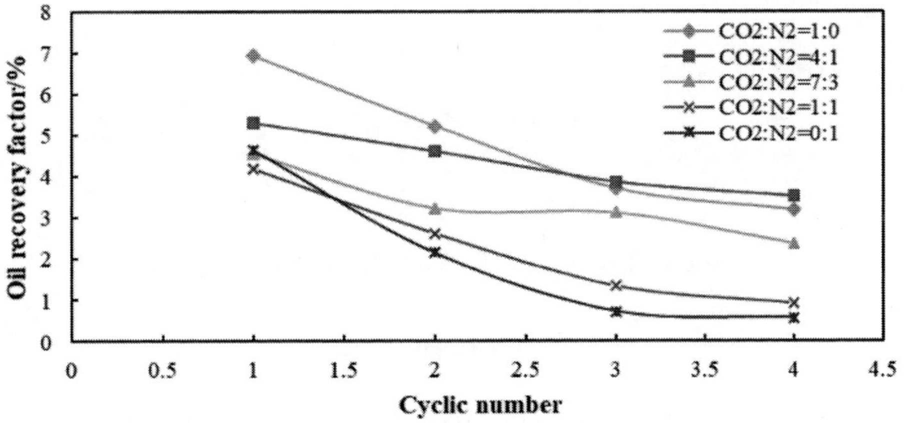

Fig.7. Oil recovery factor of each cycle for different gas injection experiments

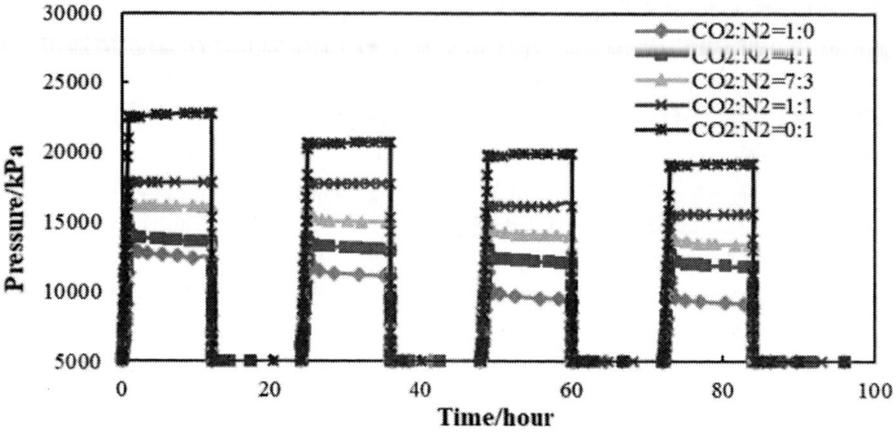

Fig.8. Pressure data of cyclic gas injection using different gas media

A pilot test results for the CO_2/N_2 mixture

Well G104-5P101 was chosen as a test well for CO_2/N_2 mixture injection at the year of 2016. The water cut of the well is 99% before the mixture injection, and CO_2/N_2 mixture was then injected into the test well with 40000 m^3 of N_2 and 300 t of CO_2 (CO_2/N_2 molar ratio = 4:1). When the well was re-produced after 1 months of soaking time, the water cut dropped from 99% to 72.3%, and the daily oil rate increased from 0.17 m^3/d to 2.95 m^3/d at the initial producing process (As shown in **Fig.9**). After 279 days of production, the CO_2/N_2 mixture recovered 275 t of heavy oil from the test well.

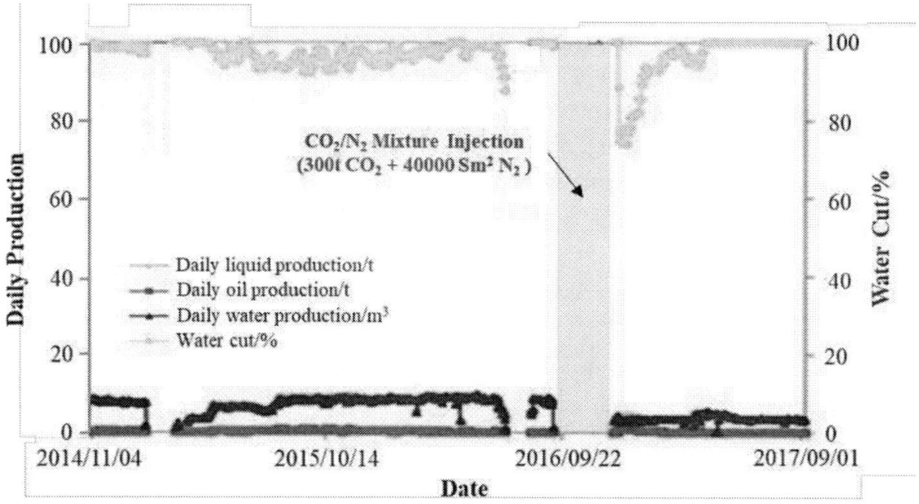

Fig.9. Test results for CO_2/N_2 mixture injection

Conclusions

A feasibility study of gas-EOR using CO_2 and N_2 mixture is evaluated to enhance the heavy oil recovery. Gas/oil interactions are firstly evaluated using a PVT analysis, then cyclic gas injection experiments using different gas media are conducted in the laboratory, and a pilot test for CO_2/N_2 mixture is also introduced in this paper. Some conclusions can be summarized as follows.

(1) The oil saturation pressure, volume factor and viscosity are highly related to CO_2/N_2 ratio of the injecting mixture. CO_2/N_2/oil interactions are usually between CO_2/oil interactions and N_2/oil interactions, and a higher CO_2/N_2 ratio can always lead to better capacities of dissolution, oil swelling and viscosity reduction.

(2) The mixture with a CO_2/N_2 moalr ratio of 4:1 can achieve an oil recovery of 17.31%. The gas/oil interactions dominated by CO_2 coupled with energy supplement provided by N_2 can effectively enhance the heavy oil recovery.

(3) A pilot test of CO_2/N_2 mixture injection is successfully conducted in a test well. An oil increment of 275 t is obtained after 40000 m^3 of N_2 and 300 t of CO_2 injection, which shows a potential EOR application of CO_2/N_2 mixture for the heavy oil.

References

1. Yu S., Li B., and Zhang Q. (1995), "Development of Jidong Oilfield with complicated fault-blocks", in *International Meeting on Petroleum Engineering*, Beijing, China.
2. Zhou H., Hao J., and Zheng J. (2006), "Development technique and practice of horizontal wells for complex fault-block reservoirs in Jidong Oilfield", in *International Oil & Gas Conference and Exhibition in China*, Beijing, China.
3. Miller J. and Jones R. (1981), "A laboratory study to determine physical characteristics of heavy oil after CO_2 saturation", in *SPE/DOE Enhanced Oil Recovery Symposium*, Tulsa, Oklahoma.
4. Mokrys I. and Bulter R. (1993), "In-situ upgrading of heavy oils and bitumen by propane deasphalting: The vapex process", in *SPE Production Operations Symposium*, Oklahoma City, Oklahoma.
5. Ivory J., Chang J., Coates R., et al. (2010), "Investigation of cyclic solvent injection process for heavy oil recovery", *Journal of Canadian Petroleum Technology*, 49 (9), pp. 22-33.
6. Shokri A. and Babadagli T. (2017), "Feasibility assessment of heavy-oil recovery by CO_2 injection after cold production with sands: Lab-to-field scale modeling considering nonequilibrium foamy oil behavior", *Applied Energy*, 205, pp. 615-625.
7. Ahadi A. and Torabi F. (2018), "Effect of light hydrocarbon solvents on the performance of CO_2-based cyclic solvent injection (CSI) in heavy oil systems", *Journal of Petroleum Science and Engineering*, 163, pp. 526-537.
8. Hudgins D., Llave F., and Chung F. (1990), "Nitrogen miscible displacement of light crude oil: A laboratory study", *SPE Reservoir Engineering*, 5 (1), pp. 100-106.
9. Sadooni M., and Zonnouri A. (2015), "The effect of nitrogen injection on production improvement in an Iranian rich gas condensate reservoir", *Petroleum Science and Technology*, 33, pp. 422-429.
10. Yuan D., Hou J., Song Z., et al. (2015), "Residual oil distribution characteristic of fractured-cavity carbonate reservoir after water flooding and enhanced oil recovery by N_2 flooding of fractured-cavity carbonate reservoir", Journal of Petroleum Science and Engineering, 129, pp. 15-22.

Heuristic Method for Determining C/N Using Inline Process Sensors

Authors:
Liu, K., University of Kentucky
Irvin, B., University of Kentucky
Frimpong, R. A., University of Kentucky
Pelgen, J. V., University of Kentucky
Nikolic, H., University of Kentucky

Built on extensive process modeling for various solvents, it has been concluded that the pH of solvent is significantly impacted by CO_2 absorbed into solution, and process temperature whilst there is negligible effect from solvent concentrations, solvent impurity and ion strength. In this presentation, inline process measurements of pH and temperature were used as inputs for a heuristic predictive model that provided a reasonable determination of carbon to nitrogen concentration ratios (C/N) in an amine solvent used in a CO_2 capture process. As CO_2 is absorbed by the solvent it becomes more acidic, pH decreases with adjusting for temperature. The C/N model was used to predict solvent carbon loading at the UKy-CAER 0.7 MWe small pilot CO_2 capture plant and a comparative analyses was performed between direct sample concentration analysis and the predictive model output. The average difference between the measured and predicted C/N was 4.6 % and the accuracy of the predicted C/N was comparable to variations among repeated sample analysis. Using the model during process operations will provide reliable, real-time feedback on process performance and solvent quality, allowing for process optimization and reducing the need for time-consuming chemical testing.

Impact of Reservoir Characterization, Field Activities, and Monitoring Results on Numerical Modeling and Forecasting of a Saline Carbon Dioxide Storage Project

Authors:
Koperna, G. J. Jr., Advanced Resources International, Inc.
Oudinot, A., Advanced Resources International

The challenge of designing monitoring, verification, and accounting (MVA) protocols for carbon capture storage and utilization (CCUS) projects is significant. Initial monitoring programs are based on cursory understandings of the subsurface, often encapsulated through numerical modeling studies that seek to forecast project performance. Further, the monitoring budget is not without limits. As such, choices are made based on the perceived value of the technology in terms of price and expected results, which are often captured in the monitoring requirements portion of the underground injection control (UIC) permit.

Selecting the most appropriate technologies is paramount in understanding the evolution of the CO2 injection project. With improved surveillance data, comes improved understanding of the evolution of the CO2 plume through history matching results and, as a result, increased confidence in the forecasting the longer-term movement of the CO2 in-situ. Making the correct choices prior to the beginning of the project can greatly improve the value proposition provided by the MVA protocols. The wrong choice, on a large-scale project, can expend millions of dollars without returning any value.

This paper will look back on a ten-year CO2 injection pilot project, supported by research dollars provided by the U.S. Department of Energy National Energy Technology Laboratory (DOE NETL) as part of their Regional Carbon Sequestration Partnerships (RCSPs), to understand the value of field-deployed monitoring data and its impact on improving the understanding of the development and movement of the CO2 plume. To do so, we will thoroughly investigate the numerical modeling alterations performed to history match the performance of the project in support of field characterization, MVA data collection, as well as field operations.

Improvement of Mobility and Conformance for CO2-EOR, a Review of Technologies and Application Conducted in China

Authors:
Long, Y., Missouri University of Science and Technology
Carbon Dioxide-Enhanced Oil Recovery (CO2-EOR), as a significant Carbon Capture and Storage (CCS) technology, exhibited enormous industrial value upon its 40 years of commercial operation. Nowadays, CO2-EOR is attracting a growing interest from both industry and academia. In accordance to the statistics reported by Al Adasani and Bai, more than 169 CO2-EOR projects have been conducted throughout the world. In China, both miscible and immiscible CO2 flooding have been implemented in Yanchang Jnlin, Daqing, Jidong, Jiangsu, and Shengli Oilfields. It was estimated that taking advantage of CO2-EOR technology, approximate 1.1 billion tons of incremental oil can be recovered from the onshore oilfields across China.

Besides its positive role in utilizing anthropogenic carbon, CO2-EOR also featured low-cost, high-efficiency, and widely applicable. The phase behavior and displacing mechanism greatly supported technique advantages. It is prone to reaching miscibility at 1500-3000 psi while nitrogen and hydrocarbon gases are holding much higher Minimum Miscible Pressure (MMP). In spite of its versatility, the sweep efficiency and reservoir conformance of CO2-EOR were still impeded due to the gas mobility, channeling problem and reservoir heterogeneity, dispersion (mixing) and gravity effects. In order to remedy these problems and furtherly improve the recovery efficiency of CO2 flooding, a new concept termed of â€œNext Generation CO2-EORâ€ was proposed. In detailed, four general technology development areas are involved in â €œNext Generation CO2-EORâ€. They are *Mobility Control Enhancement*; *Improved Flood Conformance*;*Monitoring Technology Enhancement*; *Planning and Evaluation Enhancement*. Among these, *Mobility Control Enhancement* and *How to improve the flood conformance* were challenging but achievable topics, thus substantial work have been devoted.

As a review of â€œNext Generation CO2-EORâ€ practiced in China, we tried to give a serious and comprehensive review of both laboratory experiments and field applications regarding the efforts towards an improvement of mobility and conformance for CO2-EOR in this work.

Investigating Factors Impacting Carbon Dioxide Adsorption to Shale Rock and Their Effect on Carbon Dioxide Storage Capacity

Authors:
Fakher, S., Missouri University of Science and Technology
Imqam, A., Missouri University of Science and Technology

During Carbon dioxide (CO_2) injection in shale reservoirs, a portion of the CO_2may adsorb to the rock surface, and can thus be stored in the shale reservoirs. Many factors will impact the ability of the CO_2 to adsorb to the shale rock, and the capacity of CO_2 adsorption. This research investigates the factors that can impact the CO_2 storage capacity through adsorption in shale reservoirs and the impact of each factor. The factors that impact CO_2 adsorption can be divided into three main categories, including reservoir conditions, operational factors, and rock properties. The reservoir conditions include the reservoir pressure and temperature, and fluids within the pores of the reservoir. Generally, high pressures are considered favorable for CO_2 storage since the CO_2 will be compressed and thus a larger volume will occupy the available adsorption sites. High temperatures on the other hand are not favorable due to the increase in the activity of the CO_2 molecules and thus a lower tendency for adsorption. Operational conditions includes the CO_2 injection pressure, CO_2 purity, and CO_2state, including gas, liquid, or supercritical. A lower concentration of impurities associated with the CO_2 will increase its adsorption capacity, since CO_2 exhibits physisorption, which is a multi-layer adsorption, and thus any impurities can result in a lower adsorption capacity. The rock properties are extremely important, especially the mineralogy of the rock, and the clay content. Any minerals within the shale that is soluble with CO_2 can pose some problems for long-term CO_2storage. Also, a high percentage of clays will increases the CO_2 storage capacity, since the clay structure in the shale network allows for a higher availability of adsorption sites. This research investigates some of the main factors that can impact long-term CO_2 storage in shale reservoirs and their impacts on CO_2storage capacity.

Investigating Spatial Reservoir Heterogeneity at the Shenhua CCS Demonstration Project Using CO2 Injection Simulations

Authors:
Nguyen, M. C., University of Wyoming
Zhang, Y., University of Wyoming
Wei, N., Chinese Academy of Sciences
Li, X., Chinese Academy of Sciences
Stauffer, P. H., Los Alamos National Laboratory
Saline aquifers, among other potential CO2 storage formations, have been proven to offer the highest storage capacity necessary for commercial-scale carbon capture and storage (CCS) projects. They are an ideal target for emission reduction from coal-fired power plants in the source-sink model. However, unlike hydrocarbon reservoirs which are typically well developed, these deep formations face the challenge of characterization due to a lack of data and their uncertainty. Spatial reservoir information is critical in understanding CO2 injectivity, plume dynamics, and pressure management during and after injection operations.

In this study, we combine geologic and dynamic data from the Shenhua CCS Demonstration Project in the Ordos Basin, China to provide an improved explanation for the pressure behavior and interesting injectivity at the site. We extend the work of Nguyen et al. (2017) in that a system of interconnected meandering channels are incorporated into our geologic model to account for the pore volume needed to allow high injectivity. CO2 injection simulations are carried out by varying geologic and rock/fluid parameters to generate scenarios that offer a better understanding of the subsurface spatial heterogeneity which contributed to our observed behavior. Simulation results suggest that the lateral extent of conceptual fluvial channels may not be as extensive as previously thought since the pressure front travels faster than our CO2 plume, resulting in pressure buildup at the bottom-hole level. Injection rate allocation is an important parameter in explaining the bottom-hole pressure (BHP) response at the injection well. Our models also provide a range of scenarios with the target 300,000 tons of CO2 being successfully injected into 5 layers of saline aquifers, with the sandstone Liujiagou formation receiving most of the injection.

Is Ccus Feasible? Wyoming's Triple Threat - the Cowboy Unicorn

Authors:
Carpenter, S., University of Wyoming

With the passage of 45Q there is great discussion over the â€œfinancibilityâ€ of commercial scale CCUS. While 45Q does not eliminate all the financial risks it does provide an option to allow CO_2-EOR to provide a permitting and storage â€œback stopâ€ to saline only CCUS projects.

Wyoming is special in that has a natural resource aware and likely a CCUS-willing citizenry, a positive and flexible regulatory frame work for CCUS, a supportive legislature and executive branch, and most importantly, it has the site characteristics needed to do both CO_2-EOR and saline storage. The Wyoming Integrated Test Center (ITC) located at Basin Electricâ€™s Dry Fork Station is studying options to turn CO_2 it into commercial, beneficial re-use materials â€“ this provides a CCUS project with a third option. This makes the ITC/Dry Fork Station the only facility on the planet that has the capacity, when completed, to deal with all three options for CO_2 â€“ 1) beneficial re-use, 2) CO_2-EOR, and 3) saline storage.

Imagine a place that could offer multiple coal fired power plants that could beneficially reuse CO_2 from the flue gas, AND; supply CO_2 for CO_2-EOR flooding, AND; store CO_2 in saline storage or â€œstacked storageâ€ sites. No, this is not a unicorn, but rather Wyomingâ€™s triple threat. One, if not the only place on the planet that has the capacity to deal with all three types of CO_2 from a commercial CCUS project, and consider options to ensure that there are multiple locations within Wyoming where this is possible.

This presentation will discuss the transdisciplinary approach that has resulted in Wyoming providing this world-class opportunity to advance commercial scale CCUS, current and prospective sites, potential business structures, potential realization of carbon negative oil, and the application of newly drafted International Standards for both CO_2 Storage and CO_2-EOR.

Laboratory Investigation of Sweeping Efficiency By Multiple Horizontal Wells CO_2 huff and Puff in Fault-Block Reservoirs with an Edge Aquifer

Zhixing Wang1,2,3,4, Jirui Hou1,2,3,4, Hongda Hao1,2,5,6, Fenglan Zhao1,2,6,7, Li Yan7, (1)CUP-CMG Joint Laboratory for Reservoir Simulation, China University of Petroleum (Beijing), Beijing City, China, (2)Basic Theory and Application Research Laboratory for Enhanced Oil Recovery in Low Permeability Oilfield, Key Laboratory of Tertiary Oil Recovery, CNPC, Beijing City, China, (3)Beijing Key Laboratory of Greenhouse Gas Sequestration and Oil Exploitation, Beijing, China, (4)Unconventional Petroleum Research Institute, China University of Petroleum, Beijing, China, (5)China University of Petroleum (Beijing), Beijing, China, (6)Key Laboratory for Greenhouse Gas Sequestration and Oil Exploitation in Beijing, Beijing City, China, (7)Research Institute of Enhanced Oil Recovery, China University of Petroleum (Beijing), Beijing, China,

Abstract

Water coning constrains the effective development of fault-block reservoirs by horizontal wells. CO_2 huff and puff from field cases prove to be effective in water control and oil stimulation. However, gas sweeping efficiency varies due to geological complexity and unclear oil-water interaction. Therefore, CO_2 huff and puff by multiple horizon wells needs to be optimized for a better sweep efficiency. A physical model similar to a fault-block reservoir with edge aquifer and reservoir dip was devised. Three horizontal wells were deployed according to the distance from the aquifer. CO_2 huff and puff experiments were conducted to investigate sweeping results of the selection of injecting well, combination of injection wells within the model. In addition, one numerical model corresponding to the physical one was built and numerical experiments were carried out to quantify the gas sweeping efficiency. The simulation results show that the sweeping effect decreased with the increase in the distance from the aquifer as CO_2 was injected from one horizontal well. Furthermore, when two optimized horizontal wells huffed the similar amount of CO_2, gas sweeping efficiency improved: A higher drop in water cut (34.26% to 33.53%), more oil recovery increase (15.38% to 17.59%), and a lager gas sweeping volume (33.56% to 24.66%) were obtained in the simulation experiments. It was concluded that CO_2 sweeping efficiency could be further improved within the same gas volume by horizontal wells optimization. Both physical and numerical conceptional model of fault-block reservoir was devised. And the improved CO_2 sweeping efficiency by multiple horizontal wells CO_2 huff and puff offers theoretical reference for field application.

Introduction

Water cresting from horizontal wells caused by aquifers in the bottom and at the edge of the reservoirs constrains the effective development[1-3]. As a result, water cut was hovering at high rate, and well productivity was poor. Chemical methods like conformance control by improving the permeability near the well bore calls for cautious plan for the potential environmental problem and economical cost. Gas injection is one of the oldest methods used by engineers to improve oil recovery, and its application has been increased recently. And CO_2-EOR has two major advantages: (1) additional hydrocarbon recovery that promotes energy independence and (2) CO_2 storage to reduce atmospheric emissions of CO_2.

CO_2 huff and puff has been studied for oil stimulation and coning control both in field and lab for over 60 years. D. H. Stright et al[4] invested the results of one pilot test in reservoirs with bottom and edge aquifers with numerical methods. And the simulation results showed the oil rate increased and water oil ratio (WOR) dropped. A. F.S. Palmer [5] applied the optimized huff and puff parameters to one water flooded well in West Cote Blanche Bay reservoir, and the effective time lasted over two years. M. R. Simpson [6] pointed the uniform dispersion in the formation made CO_2 huff and puff economic in both lowering water cut and boosting oil production. S. Vega Sankur et al[7] invested the PVT properties of CO_2-heavy oil system, and there was 16% in oil expansion and 45% in viscosity reduction. Haskin and Alston[8] summarized the results of 28 CO_2 huff and puff projects in Texas, and pointed out that working parameters like CO_2 slugs and soaking time should be optimized according to the oil viscosity. Charles Bardon et al[9] discussed the mechanisms of gravity drainage, CO_2 solubility in brine, imbibition, retention during the CO_2 huff and puff, and explained the process of oil saturation reduction due to interfacial reduction or wettability alteration by CO_2.

For the purpose of this paper, sweeping efficiency of CO_2 huff and puff by multiple horizontal wells in a 3D model with edge aquifers was studied on the aspect of different injection methods. Four CO_2 huff and puff experiments were conducted to investigate the CO_2 sweeping pattern. And three numerical experiments corresponding to model corresponding to the physical experiments were carried out to roughly quantify the gas and oil distribution across the radius model.

Experimental materials and methods

Materials

The crude oil used in the experiment was obtained from C2 block in the Jidong Oilfield, is the mixture of reservoir crude oil with kerosene. And the oil viscosity is 189mPa·s under the reservoir condition(60 ℃, 16.4 MPa). Brine of the edge aquifers was collected from the production well with a salinity of 937mg/L, $NaHCO_3$ type. The CO_2 was from the Jinggao Gas limited company with a purity of 99.9%.

Physical model

The 3D radius model was obtained by outcrop sand cemented by epoxy and clay minerals. The diameter of the 3D model is 400mm, and the thickness is 45mm. There are two layers with different permeabilities, $500\times10^{-3}\mu m^2$ of the upper layer with 20mm thick and $1000\times10^{-3}\mu m^2$ of the sublayer with 25mm thick. In order to reduce error from the model permeability, 5 core samples were made by the same method along with the 3D model. Core properties are listed in **Table 1**.

Table 1 Summary of core properties for the experiments of CO_2 huff and puff in the 3D models

Case NO.	Well(s) CO_2 injected	bulk volume, cm^3	pore volume, cm^3	porosity, %	Oil saturation, %
1	Well 4	4552	842	18.49	75.06
2	Well 2+ Well 4	4910	856	17.43	68.46
3	Well 3 + Well 4	4750	876	18.44	71.35
4	Well 2+Well 3 + Well 4	4505	810	17.98	64.56

There are 5 simulated wells distributed in the 3D model (**Figure 1**). Well 1 is located in the center as a monitoring well, buried 45mm deep within the model. Well 5 is on the edge of the model, severing as an edge aquifer by injecting brine under certain pressure through the ISCO pump. The other three are horizontal wells as one production unit, buried in the middle of the permeability layers with horizontal length of 8cm (Well 2 & Well 3) and 16cm (Well 4). And there are micropores (0.1mm in diameter) drilled in the surface of the simulated wells as perforation holes. The 3D model is tilted with 15° to simulate the reservoir dip during the whole process. Therefore, the three horizontal wells can be distinguished by the geological portion, Well 3 as top, Well 2 as middle, and Well 4 as bottom.

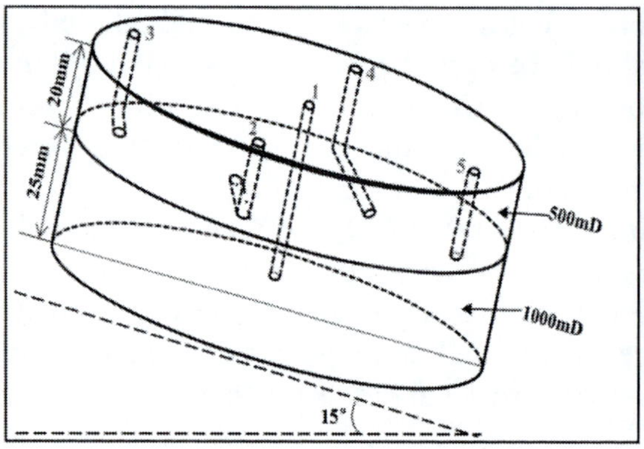

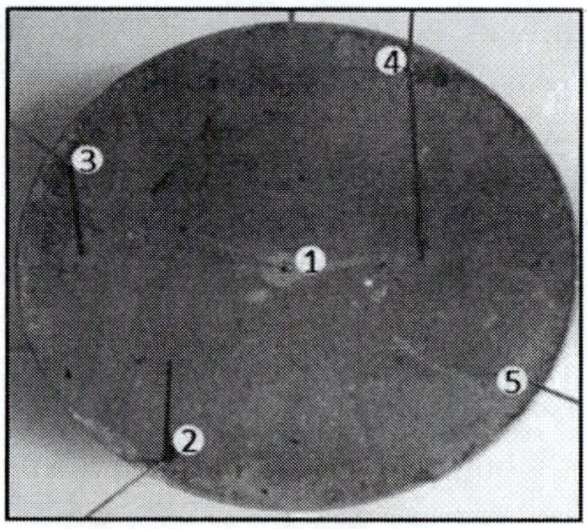

Figure 1.a Conceptional model **Figure 1.b** Actual model

Figure 1 Demonstration of conceptional and actual physical model(red lines demonstrate)

Apparatus

The CO_2 huff and puff by multiple horizontal wells schematically shown in Fig.2 was utilized in the experiments. A radius core holder specifically designed to the 3D model size with an inner diameter of 45cm diameter and depth of 10cm was used as a radius core holder for the CO_2 huff and puff experiments. This special core holder could sustain pressures up to 20MPa and temperature to 120°C. There are two high-pressure pumps (Model 100DX, Teledyne Technologies) working separately to pressurize the CO_2 (1 in Fig.2) and control the flow rate of the edge aquifer (5 in Fig.2). Pressures in the radius core holder were measured using pressure transducers (JYB-KO-H, Beijing ColliHigh sensing technology co., LTD). The produced gas and liquid were separated and measured by gas flowmeters (LF420-S, Laifeng Scientific Technology co., LTD) and graduated test tube. The aquifer injection system shown 7 in **Figure2** contains three main parts: high pressure pump, intermediate cylinder(brine) and back pressure regulator.

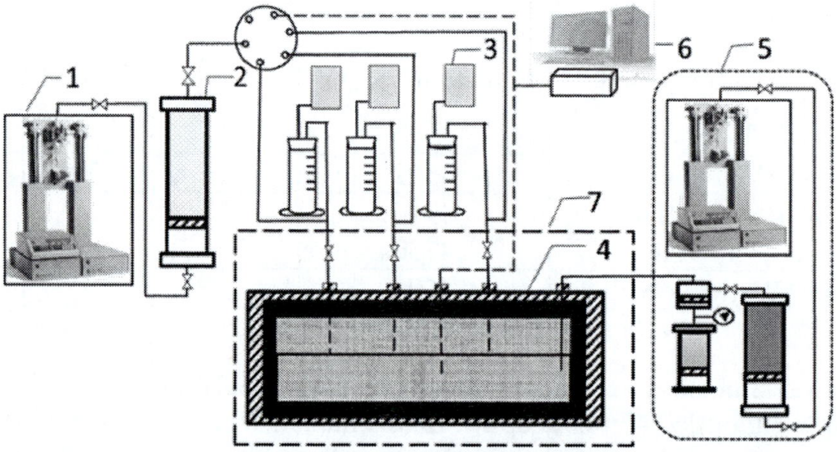

Figure 2 Experimental apparatus of CO_2 huff and puff by multiple horizontal wells

1. high-pressure pump (Model 100DX, Teledyne Technologies), 2. CO_2 cylinder, 3. Production measurement system, 4. HTHP radius core holder, 5. Edge aquifer injection system, 6. Data process system, 7. Air bath.

gas injection methods

There are four CO_2 injection strategies in this study: ① CO_2 injected from one horizontal well (the bottom, Well 4), ② CO_2 injected from two horizontal wells (the top, Well 3 and the bottom, Well 4) ③ CO_2 injected from two horizontal wells (the middle, Well 2 and the bottom, Well 4) ④ CO_2 injected from all horizontal wells. And CO_2 was injected simultaneously to the horizontal wells.

Calculation and measurement of the CO_2 volume

As introduced in the previous section, the CO_2 volume can be calculated by the density changes in a constant volume cylinder with a movable piston before and after injection at the constant temperature. CO_2 equation of state (EOS) in the density form is below,

$$PM_{CO_2} = z\rho_{CO_2}RT \tag{1}$$

The density difference before and after CO_2 can be described,

$$\Delta\rho = \rho_1 - \rho_2 = \left(\frac{P_1}{z_1} - \frac{P_2}{z_2}\right)\frac{M_{CO_2}}{RT} \tag{2}$$

The cylinder volume is 1000 cm^3 under this experiment condition, then the mass of the injected CO_2 can be calculated. Additionally, the volume under standard and experimental condition can be calculated by the CO_2 EOS in mass form (Eq. 3). Tab.3 summarized the gas injection volume under

$$PV = z\frac{m_{CO_2}}{M_{CO_2}}RT \tag{3}$$

Table 2 injected and produced CO_2 volume of different injection methods

Model NO.	temperature, K	cylineder pressure, MPa		density, g/cm^3		injected volume	
		before	after	before	after	SC, cm^3	EC, PV
1	333.15	5.75008	5.42253	0.11793	0.10912	4876.79	0.05361
2	333.15	5.85659	5.52253	0.12087	0.11177	5037.32	0.05447
3	333.15	5.85209	5.51253	0.12075	0.11151	5114.81	0.05405
4	333.15	5.84608	5.55111	0.12058	0.11254	4450.55	0.05086

*SC referred to the standard condition, 273.15K, 0.1MPa

**EC referred to the experimental condition, 333.15K, 7.5Mpa

Results and discussions

Physical Experiments

Table 2 summarized the overall results of different CO_2 injection methods. The oil recovery by water flooding from the edge aquifers were no more than 16%, more than 84% oil remained within the model. While the overall water cut from the three horizontal wells were over 90% percent, water coning in the horizontal wells constrained the well.

From the aspect of water coning control, the overall water cut decreased after the model soaked up CO_2 from different wells. Furthermore, the results differed due to CO_2 injection methods. For case 1, Well 4 in the vicinity of edge aquifer was selected for control water coning, and the water cut dropped

33.53% accordingly. However, the drop of water cut became less significant in Case 3 (CO_2 injected from Well 3) by increasing sweeping volume in high portion and Case 4 (CO_2 injected from all horizontal wells) by evenly distributed the injected CO_2. Consequently, the lowest drop was observed in Case 2, when CO_2 was injected from Well 4 and Well 2 simultaneously. For the supplemented CO_2 covered more of the water flooded area than the other injection methods.

Table 3 water control results of different CO_2 injection schemes

Case NO.	water cut, %			invasion volume*/PV	recovery factor, %	
	initial	lowest	Overall decrease		edge water	CO_2+edge water
1	93.36	59.83	33.53	0.92	15.40	15.38
2	90.93	56.77	34.16	0.87	15.90	17.49
3	94.33	64.67	29.66	0.78	15.71	16.56
4	94.16	65.08	29.08	0.76	15.62	17.34

*** Edge water invasion volume was measured by pore volume (PV) as the water cut reached 90%.**

Oil production was also stimulated as a result of CO_2 injection and the subsequent edge water flooding. Viscosity reduction and volume expansion due to CO_2 solution improved the oil mobility. Detailed increased oil recovery was demonstrated in **Figure 3**. For Well 4, located in the lower part of the model, increased oil factor declined with the increase of injection wells, for the energy supplied by CO_2 was desegregated across the model. And for Well 2, which was located in the middle of the model, the increased oil recovery maintained around 6%, because CO_2 from the lower part migrated upward and swept the area due to density difference. For Well 3 in the high portion of the model, oil recovery increased from 4.32% in Case 1 to 7.64% in Case 4. One contribution was the effect of gravity segregation during soaking period, CO_2 accumulated in the upper part of the model, thus forming a secondary gas tap. The other reason was the relatively high oil saturation near wellbore compared to the other two wells, which were experiencing severe water coning by the edge aquifer.

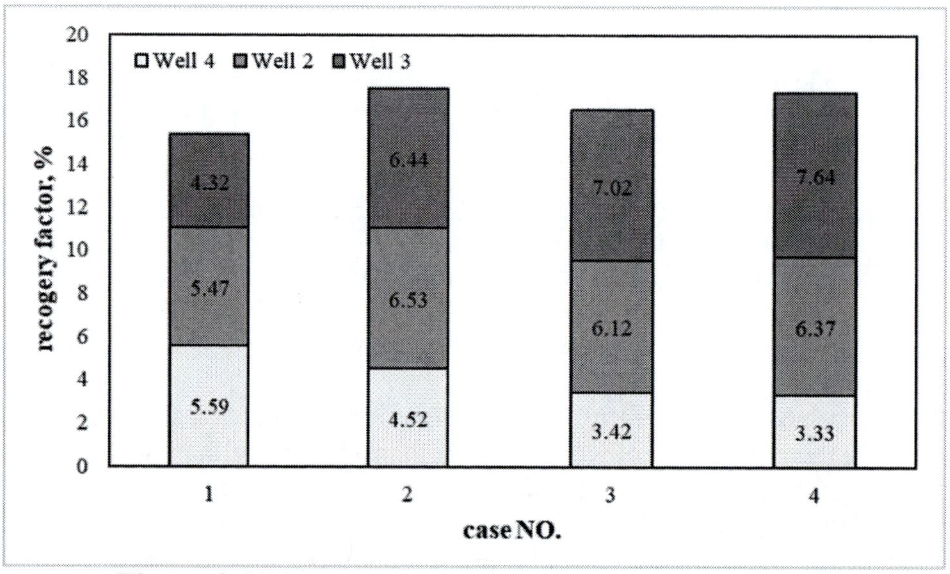

Figure 3 Oil Recovery comparison of different CO_2 methods

Numerical simulation

Gas saturation distribution after CO_2 injection and soak periods in the radius model were to compared to further present the gas sweeping efficiency of the different injection methods. Gas saturation profile of layer3 and side view after CO_2 injection were compared, which represented the gas distribution horizontally and vertically. When CO_2 was injected from Well 4, gas migrated upward near Well 3 and Well 2 (**Figure 4.a**), and downward to the edge aquifer (**Figure 4.b**). Contrary to SAGD, the gas moved to the upper position due to low density, while downward similar to SAGD since the oil density increased due to the dissolution of CO_2. Therefore, the CO_2 swept 57.91% of the layer 3 with 1002.58 cm^3 volume. When CO_2 was injected from Well 4 and Well 2, the sweeping efficiency increased to 59.64%, and the CO_2 migrated near the Well 3, which was not huff CO_2. However, the gravity drainage effect drop down since the CO_2 swept less edge aquifer areas, even though the sweeping efficiency reached 63.86%.

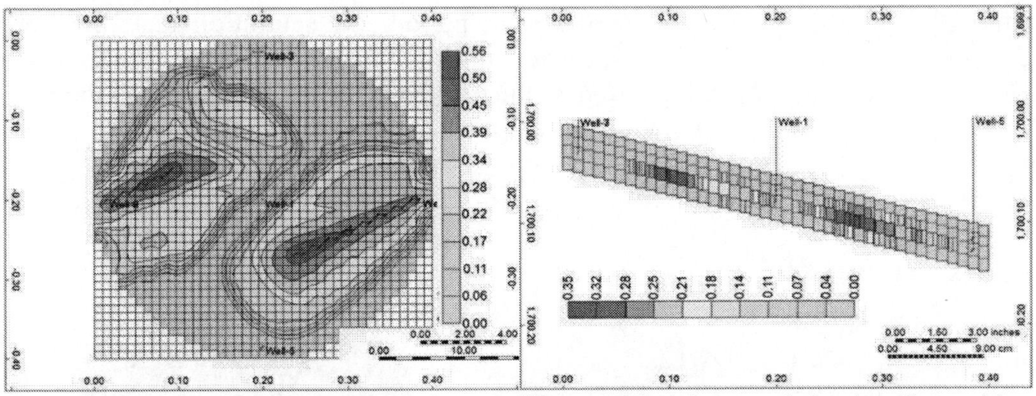

Figure 4.a After gas injection in layer3 **Figure 4.b** After gas injection from side view

Figure 4 Gas saturation distribution during soak period in gas injection from W4 and W2

Furthermore, gas sweeping efficiency of the whole radius model was compared. For case 1, namely CO_2 was injected from Well 4, the lower part of the model, the CO_2 sweeping volume was 1461.04 cm^3 with 24.66% sweeping efficiency; for case 2, when CO_2 was injected from Well 4 and Well 2 simultaneously, CO_2 sweeping volume and sweeping efficiency increased 527.03 cm^3 and 8.90% respectively compared to those of the case 1; and for case 4, when CO_2 was injected from all the three horizontal wells simultaneously, there was a less obvious increase in gas sweeping capacity. Therefore, increasing horizontal wells could achieve a higher gas sweeping volume and efficiency. Since the dissolution of CO_2 into the reservoir liquids, there was a decrease in gas saturation profiles in the model after the same soaking period. Moreover, the existence of the edge aquifer required would the maintenance of reservoir pressure to constrain the coning effect. Therefore, there should be the balance between the sweeping efficiency before and after the soak period, for Case 2, when CO_2 was injected from Well 4 and Well 2 simultaneously, there was still 1609.73 cm^3 sweeping volume across the radius model, top of all the three cases.

Table 4 Summary of sweeping volume in different gas injection patterns

Model NO.	Well(s) CO_2 injected	sweeping volume, cm^3		sweeping efficiency, %	
		after injection	after soak	after injection	after soak
1	W4	1461.04	1124.9	24.66	18.99
2	W2+ W4	1988.07	1609.73	33.56	27.17
3	W2+W3 + W4	1990.65	1576.58	33.60	26.61

Oil saturation distribution of Case 2 was studied in this part. After CO_2 was injected from Well 4 and Well 2, oil saturation varied from 12% near the well bore area to 48% in the further bore zones, corresponding to the gas saturation distribution (**Figure 4.a**), since CO_2 replaced the pore space of oil, which was displaced far from the injection well (**Figure 5.c**). More interestingly, there was a belt with high oi saturation (up to 60%) between Well 2 and Well 4(**Figure 5.a**). CO_2 could not sweep every corner within the limited volume. After the soak period, oil saturation near the horizontal wells recovered (**Figure 5.b**) to 72%, the highest across the radius model due to the gravity drainage and CO_2 solution. Moreover, oil saturation increased gradually from the edge aquifer to the horizontal wells vertically (**Figure 5.d**). The cooperation of supplied CO_2 and innate edge water redistributed oil vertically.

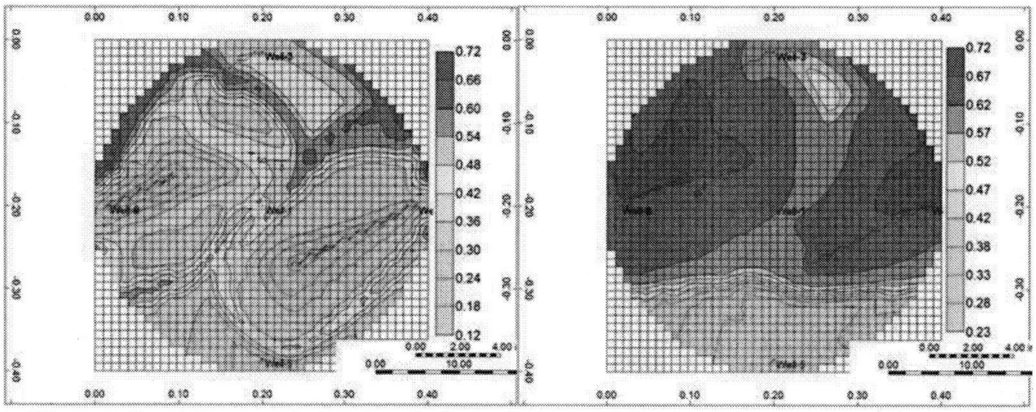

Figure 5.a After gas injection in layer3 **Figure 5.b** After soak in layer3

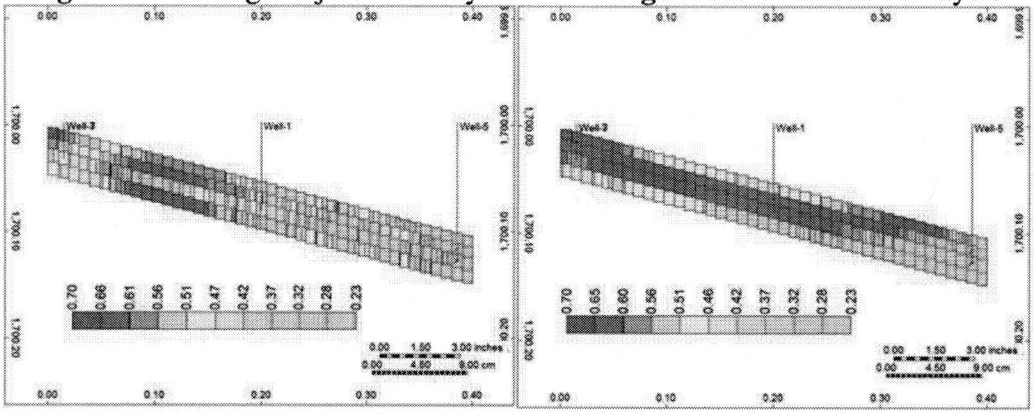

Figure 5.c After gas injection from side view **Figure 5.d** After soak from side view

Figure 5 Oil saturation distribution during soak period in gas injection from W4 and W2

Conclusion

Both physical and numerical experiments have been carried out to study the CO_2 sweeping efficiency within the 3D model with edge aquifer by different injection methods, and the optimized CO_2 injection configuration were discussed. The conclusions were as follows:

(1) CO_2 entered the water flooded pores and expanded the oil, draining away water from the wellbore. And CO_2 huff and puff by multiple horizontal wells enhanced the water drainage effect. Therefore, overall water cut could be reduced by over 20% under this condition and water cresting or coning by edge aquifer can be suppressed to some extent.

(2) The CO_2 huff and puff from multiple horizontal wells enhanced oil mobility, further more the oil recovery. A secondary gas tap formed as the CO_2 was injected from both lower and middle horizontal wells of the radius model, making a high sweeping volume and sweeping efficiency under this condition.

(3) The injection of CO_2 into the multiple horizontal wells should be distributed carefully for a better result in both conning control and oil stimulation. For the given well configuration, there should be at least two horizontal wells for CO_2 injection, one should be close to edge aquifer for coning control, and others should be located in the areas with relatively high oil saturation.

Reference

1 P. Permadi, R. L. Lee, and R. S. T. Kartoatmodjo, "Behavior of water cresting under horizontal well," SPE30374, 1995.

2 Q. Jiang and R. M. Butler, "Experimental and numerical modelling of bottom water coning to a horizontal well," Journal of Canadian Petroleum Technology, vol. 37, no. 10, pp. 82 - 91, 1998.

3 S Huang, B Zeng, F Zhao et al. Water Breakthrough Shape Description of Horizontal Wells in Bottom-Water Reservoir, Mathematical Problems in Engineering, vol. 2014, article ID 460896, .https://doi.org/10.1155/2014/460896.

4 S.G. Sayegh, B.B. Maini. Laboratory evaluation of the CO_2 huff-n-puff process for heavy oil reservoirs[J]. 1984, PETSOC-84-03-02.

5 M. Mutailipu, Y. Liu, L. Jiang et al. Measurement and estimation of CO_2–brine interfacial tension and rock wettability under CO_2 sub- and super-critical conditions[J]. Journal of Colloid and Interface Science, 2019: 534, 605-617.

6 H Li, S Zhen, D Yang. Enhanced Swelling Effect and Viscosity Reduction of Solvents-CO_2-Heavy Oil Systems [J]. SPE Journal, 2013,18 (4): 695-707.

7 D. H. Stright, Jr, K. Aziz, A. Se!tari, F. E. Starratt. Carbon dioxide injection into bottom-water, undersaturated viscous oil reservoirs [J]. Journal of Petroleum Technology, October, 1977: 1248-1258.

8 F.S, Palmer, R.W. Landry, S. Sou-Mikael. Design and Implementation of Immiscible Carbon Dioxide Displacement Projects (CO_2 Huff-Puff) in South Louisiana. SPE 15497. Presented at the 61st Annual Technical Conference and Exhibition of the society of Petroleum Engineers held in New Orleans, October 5-S, 1986.

9 Marcia Reeves Simpson. The CO_2 Huff 'n' Puff Process in a Bottomwater-Drive Reservoir. SPE-16720-PA, Journal of Petroleum Technology, July 1988: 887-893.

Microfluidics As a Tool for Rapid Testing of Chemicals for Supercritical CO2 Foams

Authors:
Gizzatov, A., Aramco Services Company: Aramco Research Center—Boston
Al Yousif, Z., Saudi Aramco
Pierobon, S., Interface Fluidics Limited
Abdel-Fattah, A., Saudi Aramco

Sequestering CO_2 in underground oil reservoirs is helping to improve recovery of oil from porous structure of the reservoir rock. However, the low density and viscosity of supercritical CO_2 (sc-CO_2) does not allow for efficiently sweeping the whole reservoir area and displacing water and oil which occupies space in the reservoir. This leads to early breakthrough of the CO_2 from producing wells and to inefficient sequestration. To improve sweep efficiency, surfactant additives are among chemicals being studied to increase viscosity of the sc-CO_2 flood as they allow production of viscous foams/microemulsions when the sc-CO_2 and aqueous phases are mixing. A large number of reservoirs have harsh conditions with high salinity brines and temperatures exceeding 90 â°C. Developing surfactants for such conditions requires testing of a large number of candidates. Conventionally used techniques, like core flooding and foam loop rheometer experiments are expensive, complex, consume hundreds of mL formulations, and can take days to perform a measurement. This is not desirable when a large number of samples need to be screened. This work presents results performed on a microfluidics device used to screen a large number of samples and to evaluate the strength of the foam in the presence of sc-CO_2. Conventional foam rheometer testing was also used to compare results. The study suggests that the microfluidics device provides the capability to perform rapid screening of chemicals with one measurement taking less than an hour. Among other advantages are the utilization of only a small volume of sample, ability to visualize flow in pore spaces under the microscope, and increased safety due to reduced size of the highly pressurized instrument. Core flooding and conventional rheometer measurements are still of great importance and should be used to further evaluate best performing chemicals that have been downselected from the larger list of formulations.

Models for Calculating Corrosion Rates in Water-Saturated and Under Saturated CO2 Systems & Water Solubility in CO2 Systems at Supercritical Conditions

Authors:
Kowta, R., Wood Inc
Erickson, D., Wood Inc
Barker, R., University of Leeds
Neville, A., University of Leeds
Hua, Y., University of Leeds
Accurately predicting internal corrosion rates in high-pressure CO_2 environments is critical to the design and operation of pipelines used for Carbon Capture and Storage (CCS) systems. In water-containing dense phase CO_2 environments, corrosion rate increases in conjunction with humidity, typically reaching a critical humidity, beyond which corrosion rates become particularly high. The presence of impurities in the CO_2 stream influence the solubility of water and hence the humidity and corrosivity of the CO_2 stream. If the CO_2 stream is under saturated and below a particular humidity level, then there can be significant cost saving on CCS projects, as this enables carbon steel pipelines to be implemented with confidence. In this work, we discuss the development of a model to calculate corrosion rates in CO_2 systems, both pure as well as those containing impurities, with under-saturated water content at supercritical conditions. For CO_2 systems with water-saturated conditions, we present corrosion rates calculated from existing in-house models. We also present experimental data used in validating the corrosion rates calculated from these models.

The model developed to calculate corrosion rates at under-saturated water conditions uses the solubility of water in the CO_2 rich phase at saturated conditions. A modified Peng-Robinson equation of state, E-PPR78, available in literature is modified to match water solubility in the CO_2 phase for CO_2 systems in the presence and absence of impurities at both supercritical conditions and on both sides of the two phase-region. We present results from the newly developed model, validating calculated water solubility in the CO_2 rich phase against experimental data available in literature over a wide range of pressures and temperatures. For pure CO_2 with water system, the error in the calculated water solubility in CO_2 rich phase is less than 4% in sub-critical region and is less than 7% in the super-critical region.

CMTC-558164-MS

Modifying Peng-Robinson Equation of State to Consider Influence of Confinement on Fluid Phase Behavior

Yilei Song[1,2], Zhaojie Song[1,2,], Jirui Hou[1,2] and Zhixing Wang[1,2], (1) Unconventional Petroleum Research Institute, China University of Petroleum, Beijing, China, (2) Beijing Key Laboratory of Greenhouse Gas Sequestration and Oil Exploitation, Beijing, China*

Abstract

The behavior of confined fluids in shale reservoirs is significantly different from that of bulk fluids due to the confinement of nanopores. The unique physicochemical properties of confined fluids invalidate the current available equations of state (EOS) used in such reservoirs. In this study, first, the molar volume term of the Peng-Robinson equation of state (PR-EOS) is modified to describe the phase behavior of fluid under confinement. Second, a new dimensionless parameter β is proposed to describe fluid adsorption in nanopores and its empirical correlation is given. Finally, based on the modified PR-EOS, a generalized analytical formula is proposed to calculate critical temperature and pressure shifts by considering fluid adsorption in the nanopores. The modified PR-EOS, new adsorption correlation, and generalized formulations have been validated to be accurate by comparing with the experimental and molecular simulation results. The modified PR-EOS can provide a consistent fluid phase behavior description over the macro-to nano-scale pore sizes.

Keywords: modified PR-EOS, nanopores, confined fluids, critical shifts

Introduction

As an unconventional resource, shale oil point to a promising future. Despite the great success of some development methods (horizontal drilling and multi-stage fracturing), many challenges remain, such as inaccurate estimates of reserves and production due to unknown fluid phase- and flow-behavior within tiny pores. The fully understand of fluid phase behavior is critical for efficient production of all reservoirs. In both conventional and unconventional reservoirs, inappropriate characterization and erroneous phase behavior models are the main reasons for the industry to face many production problems(Salahshoor et al., 2018).

The diameter of pore throat in conventional reservoirs is more than 2 μm, however, the shale reservoir is dominated by nanopores, and the diameter of pore throat tends to be 1-100nm(Nelson, 2009). As the system scale is reduced to the nanometer scale, the molecular-molecular and molecular-wall interactions are enhanced, that is, the confinement effect is significant, which leads to fundamental

* Corresponding author, songz@cup.edu.cn (Z. Song).

changes in the thermodynamic characteristics and phase equilibrium of the fluid(Kamari et al., 2018; Zhang et al., 2016; Zuo et al., 2018). Specifically, the critical properties of confined fluids in nanopores are more complex than those of bulk fluids(Islam et al., 2015; Pitakbunkate, 2014; Pitakbunkate, 2016; Tan et al., 2019; Wu et al., 2016).

It has become a consensus that spatial confinement can cause critical property shifts of fluids in nanopores, however, the extent and trend of these shifts remains controversial(Jana et al., 2009; Morishige and Shikimi, 1998; Singh and Singh, 2011; Wu et al., 2016; Zarragoicoechea and Kuz, 2004). Zarragoicoechea and Kuz (2004) utilize generalized van der Waals theory to conclude that the shifts of critical temperature and pressure of confined fluids in nanopore have the same behavior, and obtained two correlations, which is commonly accepted and widely applied(Dong et al., 2016; Li and Sheng, 2017; Mohammad et al., 2018; Zarragoicoechea and Kuz, 2004; Zhang et al., 2019; Zhang et al., 2016). However, the two correlations cannot be considered universal because they assumed that the walls are inert, this means that no fluid-walls interactions are considered in their models. Recent studies have cast doubt on this(Islam et al., 2015; Singh et al., 2009; Tan et al., 2019; Wu et al., 2016). Tan et al. (2019) found, for example, that temperature decrease was 1 order of magnitude smaller than the pressure decrease using a differential scanning calorimeter(Tan et al., 2019). Experimental investigation is the earliest method to research fluid phase behavior in nanopores. However, due to the complexity of nanopores and the limitations of equipment, experimental study is very difficult and of high uncertainty. Therefore, it is of great significance to develop an effective and reliable method for modeling and predicting confined fluid phase behavior.

In this study, first, the traditional PR-EOS is modified to describe the phase behavior of fluid under confinement by the modification of molar volume term. Second, a new dimensionless parameter β is proposed to describe fluid adsorption in nanopores and its empirical correlation is given. Finally, based on the modified PR-EOS, a generalized analytical formula is proposed to calculate critical temperature and pressure shifts by considering fluid adsorption in the nanopores.

Adsorption and effective molar volume

In the pore network, fluid molecules are usually absorbed on the pore wall and kept unmoved. In large pore sizes, compared with the bulk number of liquid molecules, the limited number of absorbed molecules can be neglected. In tight reservoirs, absorption will be significant and greatly reduce the number of molecules in the fluid that can be moved. This is more pronounced in nanopores, because the total number of molecules is limited, the absorption may limit a large proportion of fluid molecules as shown in Fig.1. The reduction number of movable fluid molecules will lead to the change of effective molar volume, which requires modification of the effective molar volume term in the traditional PR-EOS, so as to apply it to the fluid in confined nanopores.

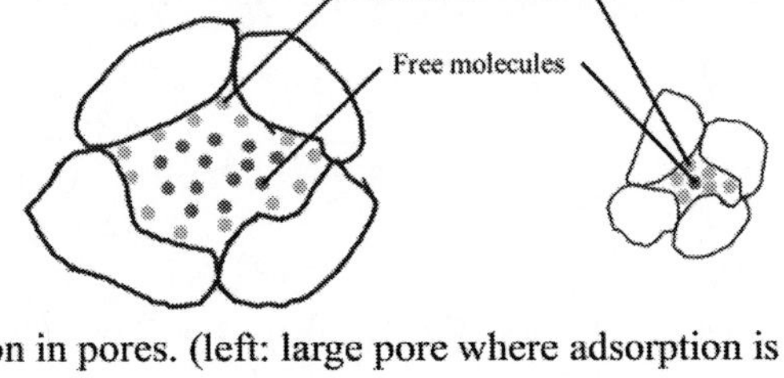

Fig. 1 Schematica of adsorption in pores. (left: large pore where adsorption is ignored, right: tiny pore where adsorption is significant)

Here, a cylindrical pore model is established to consider the effect of adsorption on effective molar volume, as shown in Fig.2. The pore model consists of two parts: Adsorption region and Free region. R_p is the pore radius, R_e is the effective radius, h is the thickness of adsorption region in radius direction, L is the length of the cylindrical pore. The effective molar volume should be the number of moles of the internal free fluid divided by the volume of the cylindrical pore.

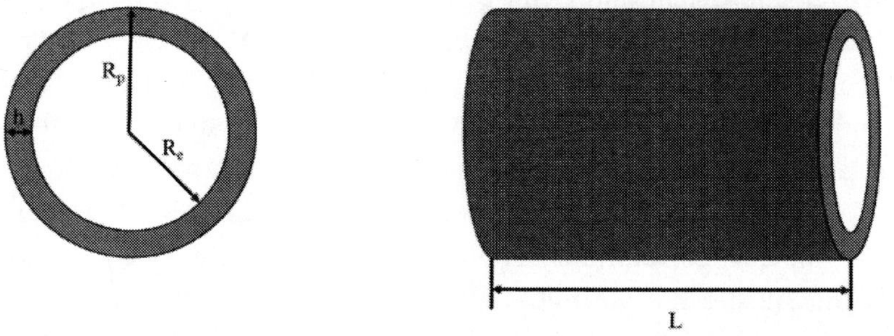

Fig. 2 Sketch map of pore model. (The adsorption region is shown in red, the center white area corresponds to free liquid.)

In this study, we assume that the fluid molecules are spherical. The adsorption region is defined as the section in which its local density deviates from the bulk, and density can be characterized by effective molecular volume coefficient ε which is defined by

$$\varepsilon = \frac{V_e}{V_{real}} \tag{1}$$

Where V_e is the effective volume of per fluid molecule, V_{real} is the real volume of per fluid molecule. The number of absorbed fluid molecule n_a can be describe as the ratio of adsorption volume (V_a) and effective volume of per absorbed fluid molecule (V_e).

$$n_a = \frac{V_a}{V_e} = \frac{\pi {R_p}^2 L - \pi {R_e}^2 L}{\varepsilon_a V_{real}} = \frac{\pi L (2R_p h - h^2)}{\varepsilon_a V_{real}} \tag{2}$$

Where ε_a is the effective molecular volume coefficient of adsorption region. The number of total fluid molecule n_t can be describe as the ratio of pore volume (V_p) and effective volume of per free fluid molecule (V_e).

$$n_t = \frac{V_p}{V_t} = \frac{\pi {R_p}^2 L}{\varepsilon_t V_{real}} \tag{3}$$

Where ε_t is the effective molecular volume coefficient without adsorption. The original molar volume V_m in PR-EOS is:

$$V_m = \frac{V_p}{n_t / N_A} \tag{4}$$

Where N_A is the Avogadro's number. The modified molar volume $V_m^{'}$ is:

$$V_m^{'} = \frac{V_p}{(n_t - n_a)/N_A}$$

(5)

It can be rewritten as:

$$V_m^{'} = \frac{n_t}{(n_t - n_a)} V_m = \frac{V_m}{1 - \frac{\varepsilon_t}{\varepsilon_a}\left(\frac{2h}{R_p} - \left(\frac{h}{R_p}\right)^2\right)}$$

(6)

We define a parameter β, which is the ratio of ε_t to ε_a and is adjustable. Thus, the modified molar volume can be desciribe as:

$$V_m^{'} = \frac{V_m}{1 - \beta\left(\frac{2h}{R_p} - \left(\frac{h}{R_p}\right)^2\right)}$$

(7)

Modifications of Peng-Robinson equation of state

The traditional Peng-Robinson equation of state (PR EOS) is as follows:

$$P = \frac{RT}{V_m - b} - \frac{a}{V_m(V_m + b) + b(V_m - b)}$$

(8)

For fluids confined in a cylindrical nanopore, substitute the original molar volume by the modified one, then the modified PR-EOS can be listed as:

$$P = \frac{RT}{V_m^{'} - b} - \frac{a}{V_m^{'}\left(V_m^{'} + b\right) + b\left(V_m^{'} - b\right)}$$

(9)

Therefore,

$$P = \frac{RT}{\frac{V_m}{\left(1-\beta\left(\frac{2h}{R_p}-\left(\frac{h}{R_p}\right)^2\right)\right)} - b} - \frac{a}{\frac{V_m}{\left(1-\beta\left(\frac{2h}{R_p}-\left(\frac{h}{R_p}\right)^2\right)\right)}\left[\frac{V_m}{\left(1-\beta\left(\frac{2h}{R_p}-\left(\frac{h}{R_p}\right)^2\right)\right)} + b\right] + b\left[\frac{V_m}{\left(1-\beta\left(\frac{2h}{R_p}-\left(\frac{h}{R_p}\right)^2\right)\right)} - b\right]} \quad (10)$$

Equation (10) is the modified PR-EOS for fluids confined in a cylindrical pore.

For any cubic equation of state, an inflection point exists at the critical point. Mathematically speaking, the first and second derivatives of pressure with respect to volume at a constant temperature are equal to zero at critical point:

$$\left(\frac{\partial P}{\partial v}\right)_{T=T_c} = \left(\frac{\partial^2 P}{\partial v^2}\right)_{T=T_c} = 0$$

(11)

Imposing Equation (11) on Equation (10) and solving for parameters a and b yields.

$$a = 0.457235\frac{R^2T_c^2}{P_c}$$

(12)

$$b = \frac{0.0777969}{\left(1-\beta\left(\frac{2h}{R_p}-\left(\frac{h}{R_p}\right)^2\right)\right)}\frac{RT_c}{P_c}$$

(13)

Then the expressions of critical pressure and critical temperature can be derived as follows:

$$P_{cm} = 0.01324\frac{a}{b^2}\left(1-\beta\left(\frac{2h}{R_p}-\left(\frac{h}{R_p}\right)^2\right)\right)^2$$

(14)

$$T_{cm} = 0.17015\frac{a}{bR}\left(1-\beta\left(\frac{2h}{R_p}-\left(\frac{h}{R_p}\right)^2\right)\right)$$

(15)

Where P_{cm} and T_{cm} are the critical pressure and critical temperature calculated from the modified PR-EOS. With the critical pressure (P_c) and critical temperature (T_c) of the traditional PR EOS, the shift of critical properties are defined as follows:

$$\Delta P_c = \frac{P_c - P_{cm}}{P_c} = 1-\left(1-\beta\left(\frac{2h}{R_p}-\left(\frac{h}{R_p}\right)^2\right)\right)^2$$

(16)

$$\Delta T_c = \frac{T_c - T_{cm}}{T_c} = \beta\left(\frac{2h}{R_p}-\left(\frac{h}{R_p}\right)^2\right)$$

(17)

Correlation for critical property shift

Equation (14-17) indicate that the molecule-pore wall interation leads to the shift of critical properties. Experimental and molecular dynamic simulation results have concluded that the critical properties decrease sharply with the decrease of poresize. Singh et al. studies the shift in the temperature as a function of the ratio of pore size to the molecule size.

In this studies, in order to eliminate the effect of molecular size, the ratio of effective molecule size (D_e) to pore size (R_p) is used in Fig.3. The effective molecular diameter is not the width or length of the molecule, but the diameter of the smallest cross section of the molecule.

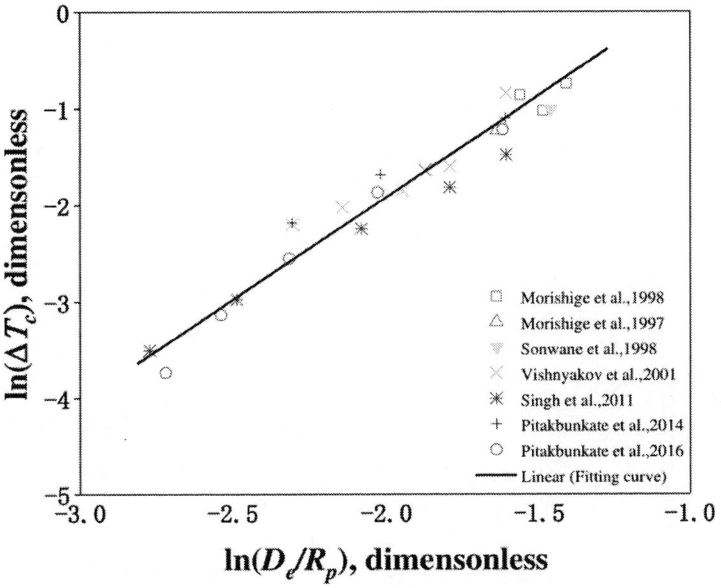

Fig. 3 Correlation of critical temperature shift versus dimensionless pore size.

There is a linear relationship between the natural logarithm of the critical temperature shift and the natural logarithm of the dimensionless pore size. The correlation of T_c-shift is quartic which is expressed as:

$$\ln(\Delta T_c) = 2.1161 \times \ln\left(\frac{D_e}{R_p}\right) + 2.3117 \quad (18)$$

The critical temperature shift can be then derived as:

$$\Delta T_c = 10.09156 \times \left(\frac{D_e}{R_p}\right)^{2.1161} \quad (19)$$

It can be seen from the above relationship and Fig. 3 that as the dimensionless pore size increases, the critical temperature change amplitude decreases, that is, as the pore size increases, the molecular wall interaction is significantly weakened.

If we substibute Equation (19) into the Equation (17), the parameter β for the modified PR-EOS can be calculated for any fluids as:

$$\beta = 10.09156 \frac{\left(D_e/R_p\right)^{2.1161}}{2\left(h/R_p\right) - \left(h/R_p\right)^2} \quad (20)$$

Where D_e set to h when single layer adsorption occurs. By determining the β-value of a particular pure component in a particular pore size, the modified PR-EOS can be solved to give the fluid properties of the component. For mixtures, similar approach can be used if reasonable mixing rules are given.

Results and discussion

We have reproduced the critical property shifts for different components with our modified PR-EOS. The results have been further validated by the experiment and molecular simulation results, as shown in Fig.4 and Fig.5.

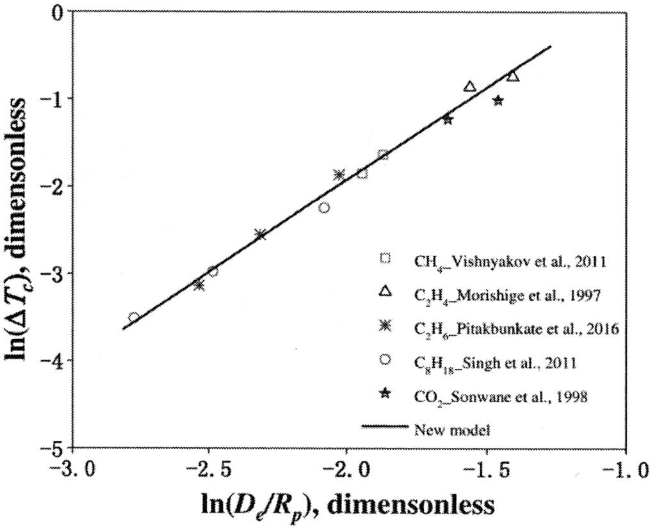

Fig. 4 Critical temperature shift from modified PR-EOS. (The symbols and lines represent, respectively, literature data and calculations.)

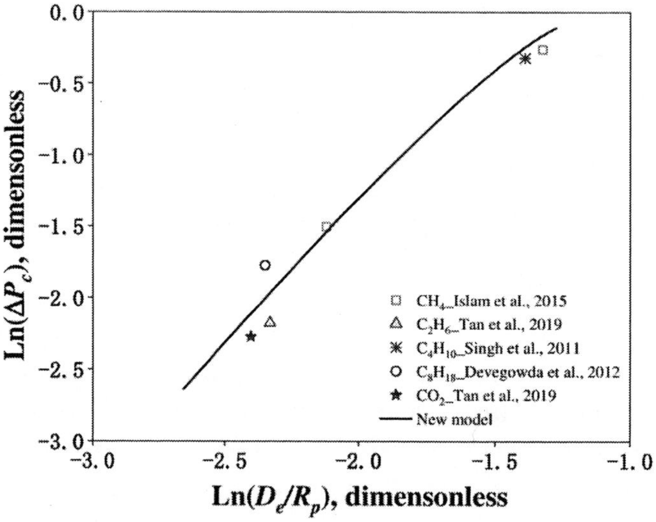

Fig. 5 Critical pressure shift from modified PR-EOS. (The symbols and lines represent, respectively, literature data and calculations.)

Fig.4 and Fig.5 shows that the relationship of critical temperature and pressure shifts and $\frac{D_e}{R_p}$ is consistent with the results of many experiment and numerical simulation. It also shows that critical pressure is more sensitive to pore proximity than the critical temperature. Due to the molecular-molecular interaction and confinement effect, adsorption layer was formed on the inner surface of the nanopore, which changed the critical properties and affected the phase equilibrium. The critical property shifts starts at R_p<100nm and increases significantly with the decrease of pore radius. In addition, the critical property shifts increases with pore radius reduction, and is more obvious at very small nanopores such as R_p=1 nm.

Conclusions

The change of fluid critical properties in nanopores has long been a consensus, but the current experimental methods cannot fully explore this phenomenon. In this work, the Peng-Robinson equation of state (PR-EOS) is modified by considering the effect of nanopore confinements on effective molar volume. Based on the modified PR-EOS and the proposed dimensionless adsorption parameters, a generalized equation is established to calculate the change of critical temperature and critical pressure with the adsorption amount of nanopores, which can accurately predict the relationship between the shifts of critical properties and dimensionless pore size. More experiments are needed to verify the performance of the modified PR-EOS on the critical pressure shifts.

Acknowledgements

The authors would like to acknowledge the China University of Petroleum (Beijing). In addition, Yilei Song wants to thank, in particular, the support from Liya Zhang and Meng Li over the past year.

Nomenclature

a = attraction parameter in PR-EOS, kPa.m^3/kmol
b = van der Waals volume, m^3/kmol
P = pressure, MPa
T = temperature, K
T_r = reduced temperature
T_c = critical temperature of the original PR-EOS, K
P_c = critical pressure of the original PR-EOS, MPa
V_m = molar volume calculated by the original PR-EOS, m^3/kmol
V_m' = molar volume calculated by the modified PR-EOS, m^3/kmol

R = gas constant, $dm^3.Pa/(mol.K)$
P_{cm} = critical pressure of the modified PR-EOS, MPa
T_{cm} = critical temperature of the modified-PR EOS, K
ΔP_c = shift of critical pressure
ΔT_c = shift of critical temperature
V_e = effective volume of per fluid molecule, m^3
V_{real} = the real volume of per fluid molecule, m^3
ε = effective molecular volume coefficient
R_p = pore radius, m
R_e = effective radius, m
D_e = effective molecular diameter, m
h = thickness of adsorption region, m
L = length of the cylindrical pore, m
n_a = number of absorbed fluid molecule
n_t = number of total fluid molecule
V_a = adsorption volume, m^3
β = adsorption parameters
Z_V = compressibility factor of vapor phase
Z_L = compressibility factor of liquid phase
δ_{ij} = binary interaction coefficient
${\varphi_V}^i$ = fugacity coefficient of component i in the vapor phase
${\varphi_L}^i$ = fugacity coefficient of component i in the liquid phase

References

Dong, X., Liu, H., Hou, J., Wu, K. and Chen, Z., 2016. Phase Equilibria of Confined Fluids in Nanopores of Tight and Shale Rocks Considering the Effect of Capillary Pressure and Adsorption Film. Industrial & Engineering Chemistry Research, 55(3): 798-811.
Islam, A.W., Patzek, T.W. and Sun, A.Y., 2015. Thermodynamics phase changes of nanopore fluids. Journal of Natural Gas Science and Engineering, 25: 134-139.
Jana, S., Singh, J.K. and Kwak, S.K., 2009. Vapor-liquid critical and interfacial properties of square-well fluids in slit pores. The Journal of Chemical Physics, 130(21): 214707.
Kamari, A., Li, L. and Sheng, J.J., 2018. Effects of rock pore sizes on the PVT properties of oil and gas-condensates in shale and tight reservoirs. Petroleum, 4(2): 148-157.
Li, L. and Sheng, J.J., 2017. Nanopore confinement effects on phase behavior and capillary pressure in a Wolfcamp shale reservoir. Journal of the Taiwan Institute of Chemical Engineers, 78: 317-328.
Morishige, K. and Shikimi, M., 1998. Adsorption hysteresis and pore critical temperature in a single cylindrical pore. The Journal of Chemical Physics, 108(18): 7821-7824.
Nelson, P.H., 2009. Pore-throat sizes in sandstones, tight sandstones, and shales. AAPG Bulletin, 93(3): 329-340.
Pitakbunkate, 2014. Effect of Confinement on PVT Properties of Hydrocarbons in Shale Reservoirs.
Pitakbunkate, 2016. Effect of Confinement on Pressure/ Volume/Temperature Properties of Hydrocarbons in Shale Reservoirs.

Salahshoor, S., Fahes, M. and Teodoriu, C., 2018. A review on the effect of confinement on phase behavior in tight formations. Journal of Natural Gas Science and Engineering, 51: 89-103.
Singh, S.K. and Singh, J.K., 2011. Effect of pore morphology on vapor–liquid phase transition and crossover behavior of critical properties from 3D to 2D. Fluid Phase Equilibria, 300(1-2): 182-187.
Singh, S.K., Sinha, A., Deo, G. and Singh, J.K., 2009. Vapor−Liquid Phase Coexistence, Critical Properties, and Surface Tension of Confined Alkanes. The Journal of Physical Chemistry C, 113(17): 7170-7180.
Tan, S.P., Qiu, X., Dejam, M. and Adidharma, H., 2019. Critical Point of Fluid Confined in Nanopores: Experimental Detection and Measurement. The Journal of Physical Chemistry C, 123(15): 9824-9830.
Wu, K., Chen, Z., Li, X. and Dong, X., 2016. Methane storage in nanoporous material at supercritical temperature over a wide range of pressures. Scientific Reports, 6(1).
Zarragoicoechea, G.J. and Kuz, V.A., 2004. Critical shift of a confined fluid in a nanopore. Fluid Phase Equilibria, 220(1): 7-9.
Zhang, K., Jia, N. and Liu, L., 2019. Generalized critical shifts of confined fluids in nanopores with adsorptions. Chemical Engineering Journal, 372: 809-814.
Zhang, Y., Lashgari, H.R., Di, Y. and Sepehrnoori, K., 2016. Capillary pressure effect on hydrocarbon phase behavior in unconventional reservoirs. SPE Low Perm Symposium 2016.
Zuo, J.Y. et al., 2018. Impact of Capillary Pressure and Nanopore Confinement on Phase Behaviors of Shale Gas and Oil. Energy & Fuels, 32(4): 4705-4714.

Appendix

Rearranging modified PR-EOS into the compressibility factor form, gives

$$Z^3 - (1-B)Z^2 + (A - 3B^2 - 2B)Z - (AB - B^2 - B^3) = 0 \quad (1)$$

Where $A = \frac{aP}{(RT)^2}$, $B = \frac{bP}{RT}\left(1 - \beta\left(\frac{2R_a}{R_t} - \left(\frac{R_a}{R_t}\right)^2\right)\right)$, Z is the compressibility factor of liquid or vapor phase.

For a mixture, the van der Waals mixing rule is used for calculating constants a and b.

$$a = \sum_i \sum_j x_i x_j (a_i a_j)^{0.5}(1 - \delta_{ij}) \quad (2)$$

$$b = \sum_{i=1}^{n} x_i b_i \quad (3)$$

Where δ_{ij} is the binary interaction coefficient of components i and j. $\delta_{ij} = \delta_{ji}$, and $\delta_{ii} = \delta_{jj} = 0$.
From the EOS shown in Equation (3), the fugacity coefficient of a pure component is

$$\ln \varphi_{pure} = -\ln[Z - B] + (Z - 1) + \frac{A}{2\sqrt{2}B} \ln\left[\frac{Z + B(1+\sqrt{2})}{Z + B(1-\sqrt{2})}\right] \quad (4)$$

For a mixture

$$\ln \varphi_V{}^i = -\ln[Z_V - B_V] + \frac{b_i}{b}(Z_V - 1) + \frac{A_V}{2\sqrt{2}B_V}\left[\frac{1}{a}\left(2\sqrt{a_i}\sum_{j=1}^{nc} z_j\sqrt{a_i}(1-\delta_{ij})\right) - \frac{b_i}{b}\right]\ln\left[\frac{Z_V + B_V(1+\sqrt{2})}{Z_V + B_V(1-\sqrt{2})}\right] \quad (5)$$

$$\ln \varphi_L{}^i = -\ln[Z_L - B_L] + \frac{b_i}{b}(Z_L - 1) + \frac{A_L}{2\sqrt{2}B_L}\left[\frac{1}{a}\left(2\sqrt{a_i}\sum_{j=1}^{nc} z_j\sqrt{a_i}(1-\delta_{ij})\right) - \frac{b_i}{b}\right]\ln\left[\frac{Z_L + B_L(1+\sqrt{2})}{Z_L + B_L(1-\sqrt{2})}\right] \quad (6)$$

where Z_V and Z_L are the compressibility factors of the vapor and liquid phases; A_i and B_i (i = V, L) are the constants calculated from the vapor pressure or liquid pressure.

$$A_V = \frac{aP^V}{(RT)^2}, \quad B_V = \frac{bP^V}{RT}\left(1 - \beta\left(\frac{2R_a}{R_t} - \left(\frac{R_a}{R_t}\right)^2\right)\right) \quad (7)$$

$$A_L = \frac{aP^V}{(RT)^2}, \quad B_L = \frac{bP^L}{RT}\left(1 - \beta\left(\frac{2R_a}{R_t} - \left(\frac{R_a}{R_t}\right)^2\right)\right) \quad (8)$$

Nano-Silica Hybrid Gel System for the Conformance Improvement of CO2-EOR

Authors:
Long, Y., Missouri University of Science and Technology

To take full advantage of CO_2-EOR, several methods related to conformance control have been discussed as an important technical suggestion. Due to the unfavorable mobility ratio and heterogeneities of reservoirs, in-situ gel treatment combined with CO_2 WAG flooding has been developed many years to solve such problem. Bulk gel consists of acrylamide based polymer was regarded as the mainstream technique which could inject into the target fracture or high permeability zones works as plugging agents to prevent the production or channeling of CO_2.

Low toxic cross-linker PEI was introduced to obtain rudimentary 3D networks through covalent bonding with acrylamide groups. The effect of major parameters included polymer, PEI concentration, molecular weight and total dissolved solids (TDS) on the performance of gelation time, strength and thermal stability were comprehensively investigated in this thesis.

The syneresis of polyacrylamide gels were frequently reported in most fields at high temperature and high content of divalent ions condition. To overcome such problems, copolymers including AMPS moieties was considered as a thermal stable polymer used in this work. Aqueous nanoparticle silica was formulated to enhance the viscosity and viscoelasticity of the hybrid gel through the inherent additional bonding effect. And the poor long time thermal stable performance in the presence of high calcium content could be removed in advance by including a powerful chelating agent EDTA.

The viscosity was measured by gel code and viscometer and the viscoelasticity was described by rheometer. The vessel test was performed to evaluate the impact of supercritical CO_2 on hybrid gels. The plugging performance on the mixed matrix/fracture media was proceeded by core flooding test.

Nuclear Process Heat for Direct Air Capture of CO2

Authors:
Lockley, A., UCL

Elevated levels of atmospheric CO_2 have encouraged consideration of Direct Air Capture (DAC) and related technologies. Most DAC approaches require very large amounts of process heat (either high or low grade) to scale the technology sufficiently to make a meaningful impact on global climate. We conduct a qualitative review of extant and postulated nuclear reactor technologies, highlighting those with particular potential for use in commonly-advocated DAC processes.

Numerical Analysis of CO_2 Utilization As Hydrates in Depleted Natural Gas Hydrate Formation

Authors:
Ahmad, S., China University of Petroleum, Beijing
Li, Y., China University of Petroleum Beijing
Li, X., China University of Petroleum Beijing
Xia, W., China University of Petroleum, Beijing
Chen, Z., China University of Petroleum, Beijing

Holding CO_2 at massive scale in enclathrated solid matter called hydrate can be perceived as one of the most reliable methods for CO_2 sequestration to mitigate devastating effect on climate change due to greenhouse gases outburst. CO_2 injection in vacated natural gas hydrate porous sediment may form hydrate under low temperature and high pressure conditions but it seems very challenging on huge scale in lengthy formations.

In this study, a dynamically coupled mass and heat transfer mathematical model is developed which elaborates the unsteady behavior of CO_2 flowing into porous media and converting itself into hydrates. The combined numerical model solution by implicit finite difference method is provided in which, CO_2 phase transition, effect of hydrate nucleation by exothermic heat release and variations of thermo-physical properties has been studied during hydrate nucleation.

The results illustrate that formation pressure distribution becomes stable at early stage of hydrate nucleation process and always remains stable afterwards but formation temperature is unable to keep stable and varies during CO_2 injection and hydrate nucleation process. Initially, temperature drops due to cold high pressure CO_2 injection since when massive hydrate growth triggers and temperature increases under the influence of exothermic heat evolution. Intermittently, it surpasses formation initial temperature before CO_2 injection initiates. The hydrate growth rate increases by increasing injection pressure in long formation and it also expands overall hydrate covered length in same induction period. The results also show that the injection pressure conditions and hydrate growth rate affect other parameters like, CO_2 velocity, CO_2 permeability, CO_2 density, CO_2 and H_2O saturation inside porous medium. In order to enhance hydrate growth rate and expand hydrate covered length, injection temperature is reduced but it did not give satisfactory outcomes. Hence, by varying pressure and temperature conditions,hydrate-bearing sediment cannot be filled with hydrates in extended formations.

Performance of Preformed Particle Gel in Supercritical CO2

Authors:
Long, Y., Missouri University of Science and Technology
Conformance problems plagued gas flooding, which can result in poor sweep efficiency and high gas oil ratio (GOR). As remediation of gas bypassing and channeling, Preformed Particle Gels (PPGs) have attracted increasing interest. Albeit PPGs were extensively studied, their actual performance in gaseous or supercritical conditions has not been reported. In this work, we deployed high-pressure vessel (HPV) to investigate the real-time performance in supercritical CO2 (scCO2) where various pressures and constant temperatures were loaded. The PPG volume was monitored and quantified through this visual instrument. Moreover, comprehensive comparisons upon particle size distribution (PSD), swelling capacity, and rheology properties were established between CO2 exposed and virgin particles. Furthermore, the differential scanning calorimetry (DSC) and scanning electron microscopy (SEM) were implemented to elucidate the scCO2 effects in respects of polymer transition and gel microstructure. It was manifest that the PPGs underwent significant dehydration in the presence of scCO2. The gel-pack shrank to 85% in volume while dewatered 66.4% in weight. The dehydration was found more pronounced at higher pressures. The particles size diminished from 5.2 to 3.7 mm whereas their swelling capacity and rheology properties were not strongly impaired. In DSC measurements, the glass transition temperature (Tg) deteriorated, denoting the plasticizing effect of scCO2 may induce a redistribution on the polymer chain. Namely, the polymer chain became more flexible and less stiff, which lead the PPGs prone to shrinking in scCO2. The SEM images illustrated a minute change in the microstructure. In conclusion, the plasticizing effect, pressure effect and pH effect played pivotal roles to account for the performance of PPGs, especially severe dehydration in scCO2. This work first reported the real-time performance of PPG in scCO2 while examined its applicability for CO2-EOR. The experimental results can facilitate the designing of PPG treatment for CO2 conformance control.

Proposal of a New Standard of CO2 Concentration in Seawater

Authors:
Sato, T., University of Tokyo
Nishimura, S., University of Tokyo
Uchimoto, K., Research Institute of Innovative Technology for the Earth
Goto, K., Environmental Technology Service Co., Ltd.
Miki, M., Japan NUS Co. Ltd.
CO2 capture and storage (CCS) is one of promising means to mitigate global warming. CCS operators who inject and store CO2 into subseabed geological formations may be legally required to monitor CO2 leakage into seawater, even though such events are very unlikely. However, the base concentration of CO2 in sweater is far from constant and it is difficult to distinguish the increase caused by a leakage. An approximate linear correlation between partial pressure of CO2 (pCO2) and dissolved oxygen saturation (DO%) is used as a standard CO2 concentration in the context of the Japanâ€™s Marine Protection Act. However, the sea off Tomakomai, Japan, where an offshore CO2 storage demonstration is now carried out, is known as a mixing area of a warm current separated from the Kuroshio Current and the Oyashio Current from the Okhotsk Sea. Therefore, the linear correlation between pCO2 and DO% scatters very much. Therefore, it is necessary to derive another standard with a certain accuracy. Here, we proposed a linear relationship between dissolved inorganic carbon (DIC) + Î±DO and temperature. In 2016, ten out of 400 monitoring data (above the upper limit of 95% confidence interval) were regarded as possible CO2 seepage off Tomakomai. The CCS operator was forced to stop CO2 injection and to seek any hint of CO2 bubble seepage in vein. This costed a tremendous amount of money and was incredibly time-consuming. If both of this method and our proposed one are used, the probability will shrink to the cases above the 98% confidence interval. In addition, we applied this standard to other seawaters to investigate its universality.

CMTC--MS579484

Recovery Potential and Mechanism Investigation of the Supercritical CO_2 EOR in the Bakken Tight Formation

Sai Wang, Juan Han, Yanbo Wang, Kegang Ling, Bao Jia, University of North Dakota;
Hongsheng Wang, Virginia Tech;
Yifu Long, Missouri University of Science and Technology

This paper was prepared for presentation at the Carbon Management Technology Conference held in Houston, Texas, USA, 17-20 July 2019.

Abstract

The low recovery of oil from the tight liquid-rich formations is still a main challenge for the tight reservoir. Thus, in order to break the chains and remove the obstacle such as the low recovery factor in the Bakken tight formation, even though the horizontal drilling and hydraulic fracturing technologies were already well applied in this field, the supercritical CO_2 flooding was proposed as an immense potential recovery method for the production improvement.

In this research, we conducted a series of CO_2 flooding experiments under various injection pressure (2500psi, 2800psi, 3000psi, 3500psi), to investigate the recovery potential of the core sample from Bakken tight formation. Also, the NMR analysis was processed of the core samples flooded with CO_2 agent under the above injection pressure variables. The result comparison demonstrates that, with the supercritical CO_2 injection pressure increase, the recovery factor gets incremental trend from 8.8% up to 33% recovery. Also, the macro pore and natural fracture system were proved to contribute more on the recovery potential. After reaching the miscible phase between the CO_2 and oil in the sample, the hydrocarbon existed in the micro pores start the contribution to the recovery potential. Thus, The CO_2 was identified as a potential recovery agent and the supercritical CO_2 EOR method was proposed as the potential recovery technology due to the high recovery factor obtained in the immiscible and miscible processes.

Introduction

The Bakken formation is currently one of the largest unconventional reservoirs in the world. According to the US Geological Survey in 2013, the oil and gas system has 7.4 billion barrels of recoverable reserves[1,2]. However, the primary recovery of the Bakken reservoir is still less than 10% even though the advanced technologies such as horizontal well and hydraulic fracturing applied[3]. Compared to its rich and dense oil reserves, Bakken reservoir development and enhanced oil recovery technology has great potential, thus the gas injection has become a domestic and foreign scholar actively explored content. Improving oil recovery is the constant pursuit of oil and gas field development[4,5]. The technology of enhancing oil recovery by gas injection brings new development methods for fulfilling this purpose. Carbon dioxide is a good candidate for the type of gas injection. Carbon dioxide injected into the formation can affect the properties of crude oil and improve oil recovery[6]. In tight Bakken reservoirs with extremely low permeability, carbon dioxide injection to develop tight oils may be the most effective method for enhanced oil recovery[7]. Previous studies have shown that the factors that contribute to the improvement of oil recovery during the contact between carbon dioxide and tight oil include reducing the viscosity of crude oil, improving the fluidity ratio of crude oil to water, reducing the interfacial tension between oil and water, volume expansion of crude oil, and light components extraction, molecular diffusion, dissolved gas flooding, mixed phase effects, and improved formation permeability, etc[8,9]. At the same time, a large amount of carbon dioxide injected into the ground can alleviate the greenhouses effect that are currently plaguing the world.

In recent years, the use of carbon dioxide to develop tight oil has been supported by many American researchers and scholars. They evaluated and analyzed the CO_2 enhanced oil recovery of tight oil injection from the perspective of experiments and numerical simulations. In terms of experiments, the tight oil core displacement experiments conducted by Gamadi[10] mainly showed that, the enhanced oil recovery factor can be improved by ranging from 33% to 85%. In terms of numerical simulation, Sorensen[11] have found that by injecting carbon dioxide, the oil production can be increased by 43% to 58% by establishing various numerical simulation models. However, the numerical models used by the above scholars in the research process generally establish a homogeneous model that assigns the same value to the mesh porosity and permeability, or a heterogeneous model that assigns different values to different small-layer porosity and permeability. The above model does not meet the actual characteristics of the Bakken reservoir, so the simulation results are often too idealistic. In terms of the choice of injection medium, carbon dioxide can be mixed with dense oil under lower pressure conditions compared with other injection gases. Carbon dioxide is easier to prepare as a relatively stable gas, and there are quite abundant carbon dioxide gas reservoirs in the ground which is easy and economical to obtain. Considering the impact of current greenhouse gases on the global climate, the environmental benefits of using carbon dioxide as an injection medium are self-evident.

Nuclear magnetic resonance (NMR) is a phenomenon in which some nuclear nucleus with a spin magnetic moment absorbs electromagnetic waves of a specific frequency under an external magnetic field, thereby changing the energy state[12]. The nuclear magnetic resonance technology is widely applied in petroleum exploration and development field recently[13]. Through nuclear magnetic resonance testing of rock samples, the physical properties and fluid parameters of the formation such as reservoir permeability, porosity, oil saturation, percentage of movable fluid and mobile water saturation can be quickly obtained, which is benefit for the division, evaluation and identification of the effective reservoirs.

After the fluid (oil or water) is contained in the pores of the tight shale reservoir, the force between the fluid molecules and the surface of the pore solids is generated, and the strength of the force is related to the pore size, morphology, mineral composition, surface properties, fluid type and viscosity, etc. When performing nuclear magnetic resonance measurement on a core sample containing fluid (oil or water), the obtained T_2 spectrum indicates the pore size distribution inside the core[14]. Within the T_2 plot, the abscissa indicates the T_2 relaxation time, and the ordinate indicates the fluid component occupancy share of different T_2 relaxation times in the core. The T_2 relaxation time is characterized by the strength of the surface of the fluid molecule and the pore solid. The fluid in the tiny pore is strongly affected by the surface of the pore solid. The T_2 relaxation time of the fluid is small, and the fluid is in a restrained or immobile state. It is called a restraint fluid or a non-movable fluid[15]. The fluid in the larger pores is weakly affected by the surface of the pore solid, the T_2 relaxation time of the fluid is large, and the fluid is in a free or movable state, called a free fluid or a movable fluid. Given the T_2 cutoff at the T_2 spectrum, the fluid in the right pore of the T_2 cutoff is a movable fluid, and the fluid in the left pore is a non-movable fluid, Through using the magnitude and distribution characteristics of the T_2 relaxation time in the nuclear magnetic resonance T_2 spectrum of the fluid in the rock sample, the state of the fluid in the pore can be analyzed.

In this research, we investigated the recovery potential of the supercritical CO_2 flooding process varied with different pressure range, which lead to the immiscible and miscible phase. The recovery factor was obtained and analysis. In order to verify the results got for the flooding process, the NMR T_2 spectrum were applied and interpreted in detail. Both of those two methods provides the strong support to the essential potential of the CO_2 EOR process.

Experimental Design

The rock samples from Middle Bakken formation (well file:16089) were obtained from the Core Laboratory at UND for the purpose of core flooding experiment analysis. The core plugs were cleaned using solvent in order to elute the organic materials in the pore and then settled into a vacuum oven over 72 hours for the purpose of drying out completely. Then, the cylindrical Bakken core plug was covered with copper sheeting in order to both form a gas-tight seal on the cylindrical wall of the sample and to apply radial confining pressure. The core plugs were evacuated and saturated with the crude oil from Middle Bakken for 15 days. In this procedure, the core plugs were evacuated for 2 hours and 3000 psi over burden pressure (OBP) was applied before starting the evacuation process. Then, the crude oil was injected at the specified low flow rate until the oil was visible at the outlet of the injection system. Keeping the oil injection operation dynamically for at least 15 days for fully saturated the core samples. In between, the core samples was released out and weighed several times and finally stop the oil injection process until the weight of the core sample reach the constant value. All the weight values, include the dry weights of all samples, the weight with the copper sheeting and the post-saturated weight of the core samples were recorded accurately, in order to obtain the accurate initial oil saturation.

The Vinci core flooding apparatus will be used for CO_2 flooding in this project (Fig. 1). The system can perform single and multiphase core flood studies at reservoir conditions. It allows the evaluation of critical parameters such as brine sensitivity, return permeability, critical flow velocity and various secondary and tertiary EOR methods, including water flooding, polymer injection, miscible and immiscible gas flooding, acid treatments and microbial flooding. Relative permeability at irreducible water saturation, residual oil saturation, displacement efficiency and incremental oil recovery after implementation of the EOR process, can be determined. The computer-controlled system is provided with a unique software that allows both manual and automated operation where all key components can be controlled including pumps, valves, video capturing and data acquisition. A test sequencer also permits automated elaborate test sequences. The core holder, air operated valves, produced fluid separator if selected and necessary plumbing are mounted in an isothermal convective air bath that has been designed to provide easy access to all main components. After the core plug is taken from core, it will be installed into the core holder and the overburden pressure will applied. Then the pump will supply the pressure source to maintain the constant CO_2 injection pressure through the core plug. The operating temperature will be set to assure that the CO_2 passing through core holder is in the super-critical phase or gas phase when the operating pressure is maintained at moderate to high pressure. The entire apparatus, except for the pumps, is contained in an air bath box to maintain a constant temperature throughout the experiment. The operation temperature was set as 40°C (104°F) to make sure the CO_2 was under supercritical state. Fig. 2 shows the schematic diagram of the core flooding system. Through the CO_2 flooding process, under a series of injection pressure (2500psi, 2800psi, 3000psi, 3500psi), the final oil recovery vs. injected pore volume from the injection and production data and plots between injected super-critical CO_2 vs. oil recovery were generated.

Fig. 1 Vinci core-flood system to be used for CO_2 injection in Bakken shale samples

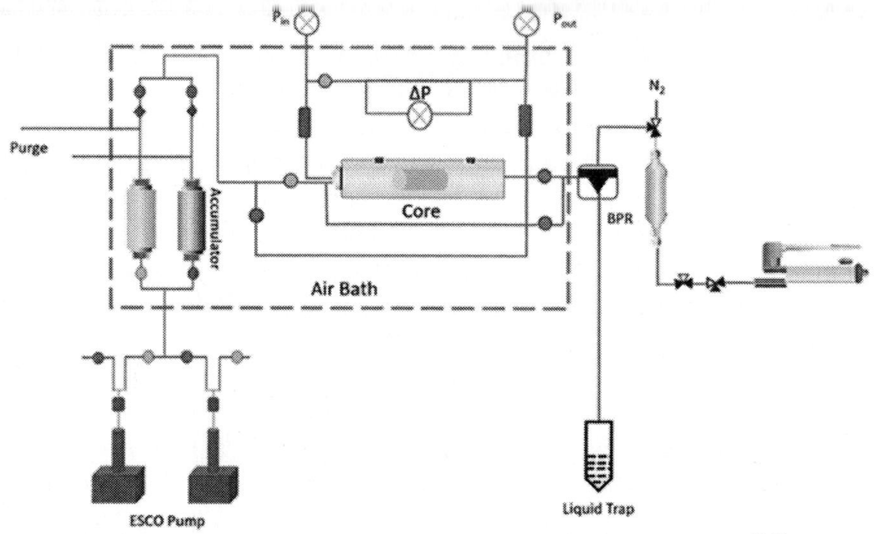

Fig. 2 Schematic diagram of the core flooding system[16].

It is critical to conduct the mechanisms investigation of the oil mobilization in the tight formation when the CO_2 enhanced oil recovery and carbon sequestration processed. After the core samples were flooded under the varied CO_2 injection pressure, the core was taken out from the core flooding system and conducted the NMR test immediately. In this work, the exposure behavior between the supercritical CO_2 and the core sample from Middle Bakken formation has been progressed experimentally by using nuclear magnetic resonance transverse relaxation time (NMR T_2) spectrum. Quantitative analysis of recovery at the pore scale was achieved. Effects of varies CO_2 injection pressure (2500psi, 2800psi, 3000psi, 3500 psi) on recovery performance have been investigated properly.

Result and Discussion

Under variable super critical CO_2 injection pressure range (2500 psi, 2800psi, 3000psi, 3500psi), the recovery potential for the core plug with the incremental injection volume of supercritical CO_2 were demonstrated in Fig.3. Based on the observation, with 6 PV of CO_2 continuously injected, the pressure was a primary factor which may give the positive effect on the recovery potential of the CO_2 flooding process. With the pressure increase, the recovery factor of core was increase from 8.9% to around 34% dramatically with the CO_2 injection pressure varied from 2500psi to 3500psi, which can be proved to be more beneficial for release the limitation of the low recovery factor in the Bakken formation. Under 2500psi, with the continuous injection of CO_2, the recovery factor was increase quickly and then trend to increase slowly then almost achieve the balance, which means not too much oil will be recovered. The main reason for this curve trend is due to the immiscible flooding process dominant. Immiscible CO_2 flooding was capable of mobilizing oil in the rock with very low permeability (0.16 mD). The injection pressure of CO_2 is below the minimum miscible pressure(MMP), which is around 2780 psi for the Middle Bakken formation, thus the CO_2 cannot penetrate into the matrix and extract the oil from the small pore. In this immiscible flooding process, the natural fracture and large pore were believed to dominate the recovery potential. Another factor may affect the recovery potential is the ability of oil extraction by supercritical CO_2.

Back to the high pressure range(2800psi, 3000psi, 3500psi), which are above the minimum miscible pressure, the recovery potential curve can be divided into three stages: 1) Before the 2PV of Supercritical CO_2 injection, the recovery factors for all pressure variables were observed rapid growth. In this stage, the produced oil was mainly from the natural fracture and pores with large size. 2) between 2 PV and 3PV of supercritical CO_2 injection, the recovery factors keep the minute increase or reach a transient equilibrium. The main reason for this phenomenon happened is that the major portion of the hydrocarbon from the natural fracture and large pore was extracted and produced, and the sweeping power of the CO_2 agent is not enough to extract the residual oil inside of the fractures and large pores, thus the maximum recovery potential of those two structures were touched. Another reason for the production delay may due to the time spend for the supercritical CO_2 penetrate into the pores with small size and start to dissolve the light components of the oil then reach the miscible phase. 3) Stage 3 is observed from the 3.5PV of CO_2 injection to the end of the injection process. In this stage, under all the injection pressure variables, the recovery factors start to increase again. The reason for this incremental trend of recovery factor is mainly due to the meso pores and the micro pores start to dominate the oil production. After contacting with the CO_2 for a certain time , the hydrocarbon fluids start to be dissolved out and then achieve the miscible phase, thus the oil begin to be produced under the pressure push energy.
We should notice that 6PV of supercritical CO_2 injection is not the end of the recovery potential increment. While considering the economic factors, the injection amount of the CO_2 should be evaluated properly.

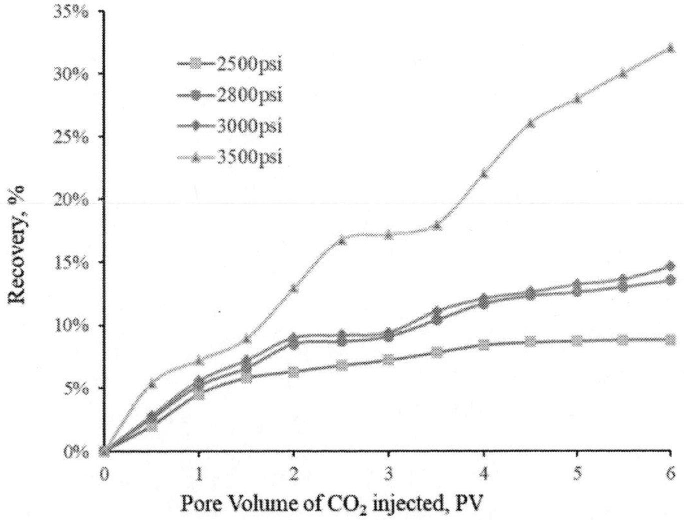

Fig. 3 Recovery potential vs. the PV of CO_2 injected, pressure range: 2500psi, 2800pis, 3000psi, 3500psi.

Fig. 4 reveals out a typical NMR T_2 spectrum of the sample from the Bakken formation. Generally, the integral area of the NMR T_2 spectra curve represents the pore fraction which hold the hydrocarbon, in other words, the oil saturation of the core sample. In the sample NMR T_2 spectra, two nanoid peak and a towering peak can be observed clearly. The towering peak can be converted and interpreted as the meso and macro pore distribution, the left weak peak stand for the micro pore existing. For the right peak in the T_2 spectrum, it is explained as the natural fracture appearance.

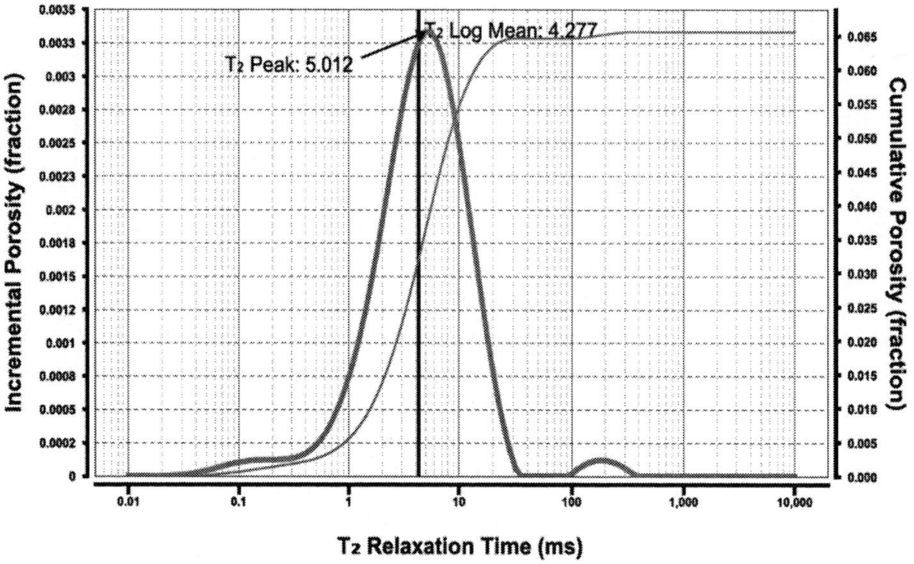

Fig. 4 Typical T_2 spectrum of the Middle Bakken Formation.

In order to interpret the NMR T_2 spectrum intuitively, the correlation conversion of NMR T_2 relaxation time and pore throat radius obtained for the N_2 gas adsorption were generated and applied[17].

T_2 can be rewritten as equation (1).

$$T_2 = Cr \qquad (1)$$

Where,

$C = 1/(\rho\ Fs)$, and C is a constant conversion coefficient (ms/μ m).

Conversion coefficient C that scales relaxation time T_2 into average pore throat radius r is determined by a method mentioned by Saidian and Prasad[18]. Through this method, the conversion coefficient C of the Bakken formation is calculated to be 2.78 ms/μm.

Fig. 5 illustrates the NMR T_2 spectra of CO_2 flooding experiment under different CO_2 injection pressure. It can be stated that the hydrocarbon in all pores can be mobilized with the CO_2 flooding pressure increase. Note that the reduction of the incremental porosity(y axis) of the pores with radius larger than 0.3μm is greater than that of pores with radius small than 0.3μm. Hence, the recovery potential of the former pores is more effective and larger than the small pore size. With the CO_2 injection, the oil existing in the pores with radius greater than 0.3μm start to diffuse to the surface of the tight matrix gradually.

Also, the towering peak of each CO_2 injection pressure variables trend to shift to the left slightly. The explanation is that the oil in the larger pore size will be recovered firstly while the hydrocarbon still remained in the smaller size pores, Thus the T_2 spectra signal which shows the total residual oil in the pores will drift to the left.

As mentioned above, the weak peak in the right side of the NMR T_2 spectra represent the recovery character of the natural fracture in the core sample. It can be seen that the left shit phenomenon also occurred during the supercritical flooding process. Also, the peak values were observed to be increased, especially under the pressure of 3500psi. It indicated that the oil in the micro pores flows to the fracture. Generally , the fracture system in the core was deemed as the main fluid flow channel. Before achieving the miscible phase, the hydrocarbon was storage in the micro pores and keep immoveable. With the CO_2 flooding pressure increase, the miscible phase will be realized and thus the oil in the micro pores will be diffused out and pushed forward to the void space which is easy to flow into. The permeability of the natural fractures in the core is larger than that of other void space, thus the natural fracture has the priority for the oil access, which mainly contribute to the peak value increase.

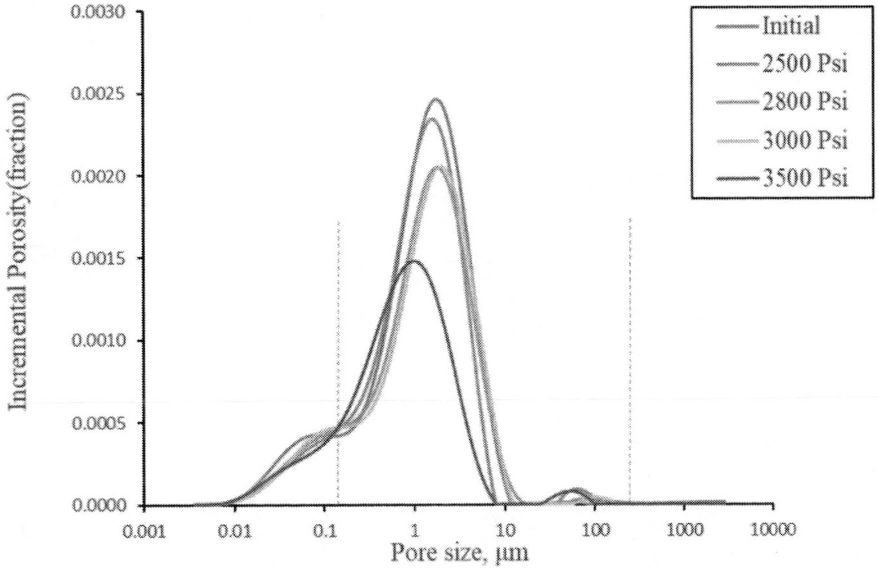

Fig. 5 NMR T_2 spectra of exposure experiment under different CO_2 injection pressure. Pressure range:2500psi, 2800psi, 3000psi, 3500psi

Table 1 Interpretation of the NMR T_2 Spectra

Initial oil saturation in the micro core	0.005187	Initial oil saturation in the macro pores and fracture		0.050784
	Pressure Range, psi			
	2500	**2800**	**3000**	**3500**
Residual oil saturation in the micro pores, fraction	0.004500	0.004238	0.004175	0.003881
Recovered oil volume from the micro pores by variable injection pressure	0.000687	0.000949	0.001012	0.001306
Recovery factor in the micro pore	0.132392	0.182911	0.195049	0.251765
Recovery factor in the total pore volume	0.012268	0.016950	0.018075	0.023330
Residual oil saturation in the macro pores and natural fractures, fraction	0.046223	0.044029	0.043873	0.032711
Recovered oil volume from the macro pores and fractures by variable injection pressure	0.004561	0.006756	0.006911	0.018073
Recovery factor in the macro pores and factures	0.089817	0.133028	0.136094	0.355881
Recovery factor in the total pore volume	0.081494	0.120701	0.123482	0.322902

Recoveries of oil in different pore size with CO_2 injection pressure variation are shown in Fig. 6. With the injection pressure increase, the recovery potential of the micro pore followed slowly increase trend. And the recovery factor from the micro pores was incremental from1.2% to 2.3% with the injection pressure increasing.

While in the macro pores and natural fractures, the recovery potential was observed to intend to effective augment with the CO_2 injection pressure increase. The recovery factor was identified as following an upward trend, which is from 8.9% to 35.6% with the supercritical CO_2 injection pressure increase.

Comparing the recovery potential from the micro pores and macro pores and natural fractures, the macro pores and natural fractures were proved to donate more contribution to the total recovery potential than that from the micro pores. The explanation may because of the different minimum mixture pressure between the two phases, the supercritical CO_2 and the oil fluid. In the macro pores and fractures with larger pore size, the MMP is relatively easy to achieve, which means the CO_2 phase and the oil phase are more comfortable to reach the miscible phase, thus the mixed fluid is easy to recovered. While in the micro pores whose pore throat is pretty small, the capillary pressure stayed in the high value. Hence the MMP is more difficulty to reach. Then the miscible phase is hard to achieve and further affect the recovery factor changed significantly.

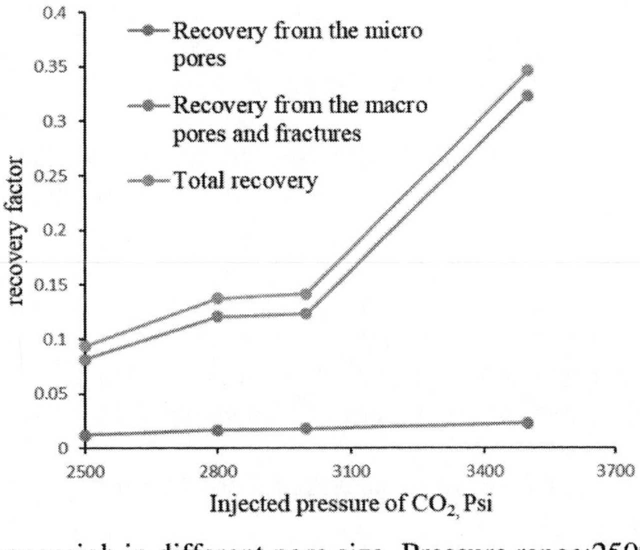

Fig. 6 Recovery weigh in different pore size. Pressure range:2500psi, 2800psi, 3000psi, 3500psi.

Fig. 7 represents the recovery potential obtained from the CO_2 flooding experiment and the NMR T_2 spectra for both of the cores. Compared with the recovery results between the flooding experiment and NMR T_2 spectra interpretation with carried injection pressure, the good match was obtain between the two curves, which demonstrated that the recovery results are reliable and accurate.

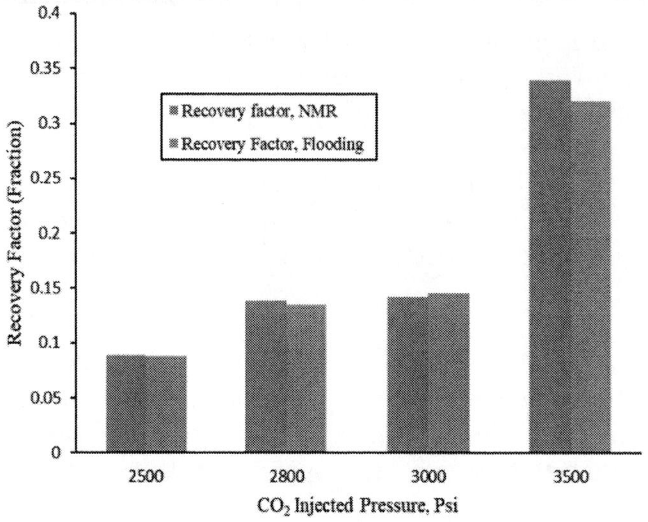

Fig. 7 Recovery factor comparison between NMR and flooding experiment. Pressure range: 2500 psi, 2800 psi, 3000 psi, 3500 psi.

Conclusions

Following conclusions can be drawn upon this study.

- A model coupling CO_2-water fractional flow in horizontal direction with CO_2-water segregation in vertical direction has been proposed. It can be used to simulate the performance of CO_2 sequestration in aquifer.
- The model indicates that CO_2 displacing water is not a piston-like displacement. CO_2 saturation at the displacement front is a function of fluid and rock properties. Behind the CO_2 front, CO_2 saturation increases gradually, and water saturation decreases accordingly until water saturation reaches irreducible water saturation.
- The propagation of CO_2 front is not linear to the injection time. its movement slows down as it moves away from injection well.
- For same radius, CO_2-water segregation causes higher CO_2 saturation at top of aquifer and low at the bottom of aquifer.
- Low vertical/horizontal permeability ratio can delay CO_2-water segregation, leads to a less severity of CO_2 overriding.

Acknowledgement

The authors are grateful to The Petroleum Engineering Department at the University of North Dakota.

Reference

[1] United States Geological Survey (USGS). USGS releases new oil and gas assessment for Bakken and Three Forks formations. USGS Webpost,

''www.usgs.gov/blogs/features/usgs_top_story/usgs-releases-new-oil-and-gasassessment-for-bakken-and-three-forks-formations/>'', May 3, 2013.

[2] Iwere F.O., Heim R.N., Cherian B.V. Numerical Simulation of Enhanced Oil Recovery in the Middle Bakken and Upper Three Forks Tight Oil Reservoirs of the Williston Basin. SPE Americas Unconventional Resources Conference, Pittsburgh, PA; June 5–7, 2012.

[3] Arshad, A., Al-Majed, A. A., Menouar, H., Muhammadain, A. M., & Mtawaa, B. (2009). Carbon Dioxide (CO_2) Miscible Flooding in Tight Oil Reservoirs: A Case Study. Society of Petroleum Engineers.

[4] Sai Wang, Jingshan San, Jianjia Yu, *et al.* (2016), A Downhole CO_2 Sensor to Monitor CO_2 Movement In-situ for Geologic Carbon Storage, International Journal of Greenhouse Gas Control.

[5] Jingshan San, Sai Wang, *et al.* (2016), Nanoparticle Stabilized CO_2 Foam used in EOR: Effect of Different Ions and Temperatures, SPE Journal, SPE-179628-PA.

[6] Jianjia Yu, Sai Wang, Ning Liu, et al, Study of Particle Structure and Hydrophobicity Effects on the Flow Behavior of Nanoparticle-Stabilized CO_2 Foam in Porous Media. SPE-169047-MS, SPE Improved Oil Recovery Symposium, 12-16 April, Tulsa, Oklahoma, USA.

[7] Jia, B., Tsau J., Barati, R. 2019. "A Review of the Current Progress of CO_2 Injection EOR and Carbon Storage in Shale Oil Reservoirs". Fuel 236: 404-427.

[8] Sai Wang, Kouqi Liu, Kegang Ling, Hongsheng Wang, Properties Evaluation of the Middle Bakken Formation Due to the CO_2 Injection. 52nd US Rock Mechanics / Geomechanics Symposium, Seattle, Washington, USA, 17–20 June 2018.

[9] Jia, B., Tsau, J., Barati, R. 2018. "Different Flow Behaviors of Low-pressure and High-pressure Carbon Dioxide in Shales". SPE Journal 23 (04): 1452-1468.

[10] Gamadi, T. D., Sheng, J. J., Soliman, M. Y., Menouar, H., Watson, M. C., & Emadibaladehi, H. (2014). An Experimental Study of Cyclic CO_2 Injection to Improve Shale Oil Recovery. Society of Petroleum Engineers.

[11] Sorensen JA, Hawthorne SA, Smith SA, Braunberger JR, Liu G, Klenner R, Botnen LS, Steadman EN, Harju JA, Doll TE. CO_2 storage and enhanced Bakken recovery research program. Subtask 1.10 final report for U.S. Department of Energy Cooperative Agreement No. DEFC26-08NT43291; May 2014. P. 79.

[12] Coates, G.R., Xiao, L., Prammer, M.G., 1999. NMR logging: principles and applications. Haliburton Energy Services.

[13] Peiqiang Zhao, Xixin Wang, Jianchao Cai, Miao Luo, Jin Zhang, Yuming Liu, Minou Rabiei, Chuncheng Li, Multifractal analysis of pore structure of Middle Bakken formation using low temperature N_2 adsorption and NMR measurements, Journal of Petroleum Science and Engineering, 2019(176):312-320.

[14] Wang H., Lun Z., Lv C., Lang D., Ji B., Luo M., Pan W., Wang R., Gong K., 2017. Measurement and Visualization of Tight Rock Exposed to CO_2 Using NMR Relaxometry and MRI. *Scientific Reports,* V7, Article number: 44354.

[15] Milad Saidian, Manika Prasad, 2015. Effect of mineralogy on nuclear magnetic resonance surface relaxivity: A case study of Middle Bakken and Three Forks formations, Fuel, V161: 197-206,

[16] Wang, S.; Liu, K.; Han, J.; Ling, K.; Wang, H.; Jia, B. Investigation of Properties Alternation during Super-Critical CO_2 Injection in Shale. Appl. Sci. 2019, 9, 1686.

[17] Francisco B., Ricardo L., Andre S., Giovanna C., Rodrigo B., 2017. Estimating the Surface Relaxivity as a Function of Pore Size from NMR T_2 Distributions and Micro-tomographic Images,Computers & Geosciences,V106: 200-208.

[18] Saidian, M. & Prasad, M. Effect of Mineralogy on Nuclear Magnetic Resonance Surface Relaxivity: A Case Study of Middle Bakken and Three Forks Formations. *Fuel* **161,** 197–206 (2015).

Research on Well Integrity during CO2 Geological Utilization and Storage Based on CT Scanning Technique

Authors:
Zhang, J., CNPC Engineering Technology R&D Company Limited
Mu, L., China University of Petroleum

The ultra-deep formations of Tarim Basin is characterized by ultra-high temperature, high pressure, and ultra-thick composite salt-gypsum formations, so it will bring great challenge to well integrity during CO_2 sequestration. Therefore it is necessary to evaluate the impact of temperature decrease and salt rock creep on injection well integrity.

Firstly, the numerical simulation method was used to analyze the influence of temperature drop and salt rock creep on the integrity of the wellbore during CO_2sequestration. The actual stress of the wellbore cement sheath was equivalent to the simulated wellbore cement by the stress equivalent method. Secondly, CT scanning technology was used to detect the internal structure of the cement sheath, revealing its failure mechanism. Lastly, with the help of GMI and WELLCAT software, the optimization of design parameters during CO_2sequestration is realized.

The results show that under some circumstances the micro-annulus between casing and cement sheath can be changed by salt creep or self-healed. The mechanism of casing damage in salt layers was preliminarily revealed. CT scans were performed on the cement sheath before and after the experiment. By comparing the gap volume changes of the casing cement sheath, it was found that there was a micro-annulus between the casing cement sheath. The mechanism of casing damage in salt layers was preliminarily revealed during CO_2sequestration. It is indicated that casing damage is mainly caused by the non-uniform load induced by the creep of salt-gypsum layers ,the low temperature ,and the poor cementing quality. Finally, a series of measures were proposed to prevent casing damage during CO_2 sequestration in the areas with a salt formation.

This new method can effectively reduce cost of casing material selection and make for guaranteeing oilfield development efficiency and CO_2 sequestration while maintaining wellbore integrity and controllable well control safety.

Sealing Wellbores at the End of Their Lifecycle to Restore Subsurface Seal Integrity and Enable Safe Ccus

Authors:
Radonjic, M., Oklahoma State University
Li, Y., OK state
Vissa, V., OK state
As we approach 2020, there are more wellbores that require abandonment than wellbores waiting to be drilled. The USA alone has ~1.7 Million wellbores with an API number indicating that they are not permanently plugged, therefore will be in need of Plugging and Abandonment (P&A) at some point in the future. Some of these wellbores are located in fragile ecosystems, such as the Gulf of Mexico, where it is estimated that 9,000 idle wellbores are waiting permanent plugging and abandonment. The numbers of future P&As suggest requirement for an urgent improvement of current technology as many governments are preparing standards that require assurance for zero leakage over thousands of years.

This paper/talk will present knowledge gaps we identified during literature review and the data obtained during first two of the ongoing project , under NASEM-GRP funding.

Extensive literature review points to the following issues:

- Portland-cement based materials, with their pH ~13, are geochemically incompatible with the reservoir conditions in the GoM.
- Weak interfaces between and contaminations with drilling fluids in wellbore structures present high risk for HC leakage, which currently cannot be mitigated successfully.
- Placement and accessibility of wellbore in offshore environment adds to technological complexity and increases the risk of leakage.
- Monitoring and verification is currently not available for long-term assessment offshore.

Some of our preliminary data suggests the following:

- Plugging materials are impacted by contaminations form drilling fluids and other wellbore materials, such as fragment of dehydrated clays, oily rock fragments.
- Casing corrosion might lead to deterioration of primary wellbore cement, which can be difficult to locate and remediate prior to plugging.
- Geomimicry is a potential path forward in designing and developing barrier materials capable of serving over thousands of year.

State of the Art for Conformance Control in CO2 EOR: A Comprehensive Review of Research and Field Applications Related to Gel Remediating Bypassed Zone

Authors:
Long, Y., Missouri University of Science and Technology

It was estimated that the implementation of Carbon Dioxide-Enhanced Oil Recovery (CO2-EOR) contributed to reducing 10-20 billion tons of Greenhouse Gas (GHG) emission in U.S. As a significant Carbon Capture and Storage (CCS) technologies, CO2-EOR possessed enormous industrial value with its 40 years of commercial operational experience. The increasing number of miscible and immiscible CO2 projects reflected industryâ€™s growing interest for CO2-EOR. It was reported that the number of miscible CO2 projects boosted to 127 from 103 projects in 2010, holding the maximum EOR practice in U.S.

There are several reasons accounting for this tendency. Besides its positive role in utilizing anthropogenic carbon, CO2-EOR also featured low-cost, high-efficiency, and widely applicable. In spite of its versatility, the incremental oil recovery of CO2-EOR is still impeded, owing to the gas mobility, channeling problem and reservoir heterogeneity, dispersion (mixing) and gravity effects. Unfavorable mobility of gas will result in an early breakthrough, therein lead to a poor volumetric sweep. Moreover, the existence of fractures, channels or high-permeability streaks in reservoir can significantly exacerbate the deficiency of CO2 flooding.

To remedy above problems and furtherly improve recovery efficiency of CO2 flooding, a new concept termed of â€œNext Generation CO2-EORâ€ was proposed. Accordingly, four general technology development areas involved in â€œNext Generation CO2-EORâ€. They are *Mobility Control Enhancement*;*Improved Flood Conformance*; *Monitoring Technology Enhancement*; *Planning and Evaluation Enhancement* . Among these, *how to improve the flood conformance* was a challenging but achievable topic, thus a large number of efforts were conducted. This paper tried to give a serious and comprehensive review of both laboratory experiments and field applications related to conformance improvement in CO2-EOR.

Study on Dynamic Phase Behavior of Oil and Gas System during CO2 Flooding

Authors:
Chen, H., China University of Petroleum

CO_2 flooding is one of the most promising EOR methods. Due to the complicated compositional change during the displacement process, the effect of CO_2 injection on phase behavior of oil and gas system is still not clear so far. It can great limits the forecast of development effect and composition optimization of injection gas. In this paper, taken an offshore oilfield as an example, based on compositional grouping, PVT test and slim tube test, phase behavior of oil and gas system during impure CO_2 flooding was studied by PVT Sim 20. Results show that under reservoir conditions, the thickness of oil layer can greatly influence the composition distribution of the reservoir fluids. Thus, the phase behavior can be significantly different. For the initial stage of CO_2 flooding, two phase region of the phase diagram continuously extend. The cricondenbar increases obviously and the cricondentherm decreases. Critical point moves towards top left. With the CO_2flooding proceeds, the two phase region begins to shrink when the gas oil ratio achieves 2:1. The cricondenbar increases slightly and the cricondentherm decreases significantly. Moreover, with the mixing of contaminated gases, both the two phase region enlarges and the cricondenbar increases greatly. However, the cricondentherm changes slightly.

Study on Operation Strategy of Aquifer Underground Gas Storage Using CO2 As Cushion Gas

Authors:
Mu, L., China University of Petroleum - Beijing
Liao, X., China University of Petroleum - Beijing
Zhao, X., China University of Petroleum - Beijing
Zou, J., China University of Petroleum - Beijing
Li, R., China University of Petroleum - Beijing
Zhang, J., CNPC Engineering Technology R&D Company Limited

The cushion gas is indispensable to maintain the reservoir pressure, prevent the water intrusion and ensure the stability of underground gas storage (UGS). However, it is inevitable that the cushion will mix with the working gas during the operation of gas storage, which reduces the thermal enthalpy and even changes the dynamics of the reservoir. It is important to research how to control the injection of CO_2 to operate the UGS efficiently.

Based on the seepage theory of two-phase, this paper establishes a two-phase three-dimensional seepage model for UGS considering the effect of the dissolution of CO_2 in water and the diffusion between the gases. The optimal operation strategy for aquifer UGS is determined through the numerical simulation of the factors that have an impact on the process of injection and production.

It can be discovered that CO_2 and the mixture with a large amount of CO_2 have much higher density and viscosity than CH_4, since CO_2 is in a supercritical state within the operation pressure and temperature of the UGS. This is inclined to restrict the mixing of the cushion gas and the working gas. The CO_2 volume suffers a great loss due to the dissolution in the aquifer, which is more likely to weaken the protection of the cushion gas. It can be seen from the variations of the width of the mixed zone that the increase of pressure, permeability the proportion of the cushion gas will widen the mixing zone. Meanwhile, the content of CO_2 in the production gas increased observably, leading to a reduction in thermal enthalpy of the production gas. Comparing the two common injection mode of the cushion gas, the result shows that the cushion gas injected through peripheral injection well has less effect.

The Applicable Limits of the High Strength Gel System to the Tight Sandstone Fractured Reservoir in CO2 Flooding

Authors:
Zhao, F., China University of Petroleum(beijing)
Xu, T., Exploration and Development Research Institute of Sinopec
Hou, J., China University of Petroleum (Beijing)
Song, L., China University of Petroleum
Lu, G., China University of Petroleum(beijing)
Feng, H., China University of Petroleum(beijing)
The matrix permeability of the tight sandstone reservoir is very low, which causes difficulty to inject water, the CO_2 flooding is applied widely to these reservoirs for the better oil displacement and low injection pressure, but natural fractures develop as well as artificial fractures exist in the tight reservoir, makes gas-channeling along the fractures severe when proceeding a CO_2 flooding process, blocking the fractures is necessary for displacing the residual oil trapped in the matrix. The self-developed tight sandstone fractured cores which can simulate different conditions of fracture width and matrix permeability had been used to evaluate the blocking effect of the high strength starch gel system, the mechanism that the high strength gel system blocked the fractures and activated the residual oil in matrix was analyzed by comparing the pressure when CO_2 broke through and the incremental recovery after blocking in CO_2 flooding experiments, then the applicable limits that the high strength starch gel system improved the oil displacement of CO_2 flooding in the tight sandstone fractured reservoir had also been discussed. The results showed that the high strength starch gel system could be used to block the fractures, and the blocking effect and the incremental recovery were the best when the fracture width was 0.42mm, however both effects became worse when the fracture width decreased to 0.08mm or increased to 0.65mm because of the injection limitation or the reduction of blocking effect; furthermore, the residual oil in the matrix could be activated effectively after the high strength starch gel blocked the fractures even if the matrix permeability reduced to 0.1mD, the research results can provide reference to the design and optimization of project when proceeding a CO_2 flooding process in the tight sandstone reservoir.

The Carbon Dioxide Removal Potential of Liquid Air Energy Storage: A High-Level Technical and Economic Appraisal

Authors:
Lockley, A., UCL
Liquid Air Energy Storage (LAES) is an energy-storage technology, currently operating at pilot plant scale. It relies on the compression, cooling and subsequent liquefaction of ambient air, as a means of storing energy. Subsequent reheating of the air (using ambient, stored, or waste heat) allows re-vapourisation of the air at significant pressure â€" to power a gas turbine or reciprocating engine , effecting energy recovery. Inherent in the liquefaction process is a need to remove water and CO2 , which would otherwise pose materials handling issues at cryogenic temperatures. This CO2 can potentially be recovered and stored (e.g. geologically), thus preventing it acting as a greenhouse gas. This Carbon Capture and Storage (CCS) add-on renders LAES a form of Carbon Dioxide Removal (CDR) , albeit as a by-product. Compressed Air Energy Storage (CAES), with additional cooling offers potentially-comparable CDR potential. Our crude numerical analysis finds that neither LAES or CAES has the capacity to conduct full planetary-scale CDR on centurial timescales. However, CDR by-products from the industry are capable of providing a trillion-dollar scale (USD) revenue stream this century to the LAES industry â€" and a large (but less certain) revenue stream to the CAES industry. This could potentially be instrumental in effecting a switch to LAES on costs grounds, as well as substantially modifying the scaling and placement of LAES or CAES plants.

The Development of CO_2 Plume in CO_2 Sequestration in the Aquifer

Hao Fu[1], Yifu Long[2], Sai Wang[1], Yanbo Wang[1], Peng Yu[3] and Kegang Ling[1], (1) University of North Dakota, Grand Forks, ND, USA, (2) Missouri University of Science and Technology, Rolla, MO, USA, (3) Beibu Gulf University, Qinzhou, Guangxi, China

Abstract

Geological carbon sequestration through injecting large-scale carbon dioxide (CO_2) into the deep saline aquifers represents a long-term storage of CO_2. In the CO_2 sequestration process, the injected CO_2 is displacing water from the injection point and is expected to remain in the reservoir. Due to the nature of one phase displacing another phase in porous media, it is noted that different water saturation exists in the CO_2 plume during the displacement. Water distribution in the plume will affect the size of the plume subsurface. Furthermore, the gravitational segregation between CO_2 and water will cause overriding-tonguing during the injection and impact the shape of plume. To better understand the CO_2 movement underground and development of CO_2 plume, it is necessary to take the two-phase flow and gravity force effects into account when evaluating CO_2 displacing water. The displacement of water by injecting CO_2 is not a piston-like process in aquifer. Because water is the wetting phase and CO_2 is the non-wetting phase when two phases flow in reservoir, water occupies the surface of matrix and small pores while CO_2 resides in large pores and centers of pores. As a result, various water saturations distribute behind CO_2 front during the displacement. The distribution is a function of fluid and rock properties, fluid-rock interaction, and injection operation. In this study, these factors are considered when developing new models to predict CO_2 plume evolution during injection. Mass conservation, multiphase flow, and equation-of-states are applied in the derivation of the models, which guarantees a rigorous approach in the investigation. The modeling results indicate that CO_2 does not displace water completely away from the plume. The shape of the CO_2 front is controlled by the relative permeability of two phases and capillary pressure. Water saturation profile from CO_2 injecting point to the displacement front shows that water saturation behind the CO_2 front increases outwardly, and the change in saturation is non-linear. The injection rate impacts the sharpness of the CO_2 front, thus leads to different gas plume sizes for same injection volume. The outward movement of the CO_2 front decelerates as injection time goes on. The research illustrates that injection experiences two stages: transient and steady-state, in which the displacement behavior and the development of gas plume vary. Although the duration of transient stage is dictated by size of aquifer and is relatively short comparing with steady-state stage, its influence on the development of CO_2 plume cannot be neglected when selecting gas compressor horsepower and determining injection rate.

Introduction

Greenhouse effect is one of the processes that cause the arguable global warming in recent years. When the energy from the Sun reaches the Earth, part of it is reflected back to the space while the rest is absorbed and reradiated by the Earth's atmosphere, which consists of greenhouse gases. CO_2 is one of the greenhouse gases that contribute to the trapping of heat that radiates from Earth. Greenhouse gases in the Earth's atmosphere are mainly CO_2, methane, nitrous oxide, hydrofluorocarbons, sulfur hexafluoride, fluorinated gases, ozone (O_3), and water vapor, etc. Table 1 (Haynes, 2016) shows the concentration of different gas in atmosphere. It is noted that the contribution of each greenhouse gas to the absorption of heat depends on its concentration and its global warming potential (GWP). GWP is used to reflect how long a greenhouse gas remains in the atmosphere, on average, and how strongly it absorbs energy. Gases with a higher GWP absorb more energy, per pound, than gases with a lower GWP, and thus contribute more to warming Earth. GWP of greenhouse gases is listed in Table 2 (U.S. EPA, 2019). Although the global warming potential of CO_2 is not the highest, its effect on global warming is significant due to the relatively high concentration and the increasing emission volume during last several decades. Many researches indicate that anthropogenic greenhouse gases emitting to the atmosphere are the main cause of global warming. Fig. 1 illustrates that the emission of CO_2 has increased dramatically since industrial revolution (Ritchie and Roser, 2019). Fig. 2 is the corresponding change of CO_2 concentration in atmosphere as a result of increase in CO_2 emission. CO_2 emission is from burning fossil fuels, solid waste, trees and other biological materials. Chemical reactions also emit CO_2.

Table 1. Major constituents of dry air, by volume (Haynes, 2016)

Gas		**Volume**$^{(A)}$	
Name	Formula	in ppmv$^{(B)}$	mole %
Nitrogen	N_2	780,840	78.084
Oxygen	O_2	209,460	20.946
Argon	Ar	9,340	0.9340
Carbon dioxide	CO_2	413.32	0.041332
Neon	Ne	18.18	0.001818
Helium	He	5.24	0.000524
Methane	CH_4	1.87	0.000187
Krypton	Kr	1.14	0.000114
Not included in above dry atmosphere:			
Water vapor$^{(C)}$	H_2O	0–30,000$^{(D)}$	0–3%$^{(D)}$

notes:
$^{(A)}$ Volume fraction is equal to mole fraction for ideal gas only,
$^{(B)}$ ppmv: parts per million by volume
$^{(C)}$ Water vapor is about 0.25% *by mass* over full atmosphere
$^{(D)}$ Water vapor strongly varies locally (Wallace et al., 2006)

Table 2. Global Warming Potentials and Atmospheric Lifetimes (U.S. EPA, 2019)

Gas	**Atmospheric Lifetime (years)**	**GWP**[a]
Carbon dioxide (CO_2)	50-200	1
Methane (CH_4)[b]	12±3	21
Nitrous oxide (N_2O)	120	310
HFC-23	264	11,700
HFC-32	5.6	650

HFC-125		32.6	2,800
HFC-134a		14.6	1,300
HFC-143a		48.3	3,800
HFC-152a		1.5	140
HFC-227ea		36.5	2,900
HFC-236fa		209	6,300
HFC-4310mee		17.1	1,300
CF_4		50,000	6,500
C_2F_6		10,000	9,200
C_4F_{10}		2,600	7,000
C_6F_{14}		3,200	7,400
SF_6		3,200	23,900

[a] 100 year time horizon

[b] The methane GWP includes the direct effects and those indirect effects due to the production of tropospheric ozone and stratospheric water vapor. The indirect effect due to the production of CO_2 is not included.

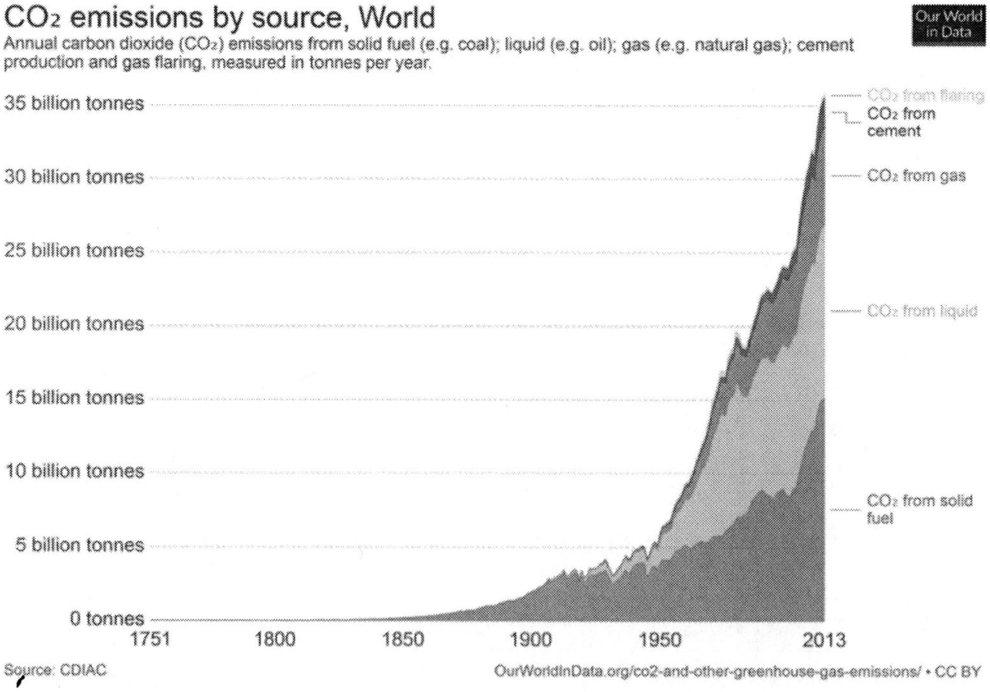

Figure 1. Global emission of CO_2 has increased significantly since industrial evolution (Ritchie and Roser, 2019)

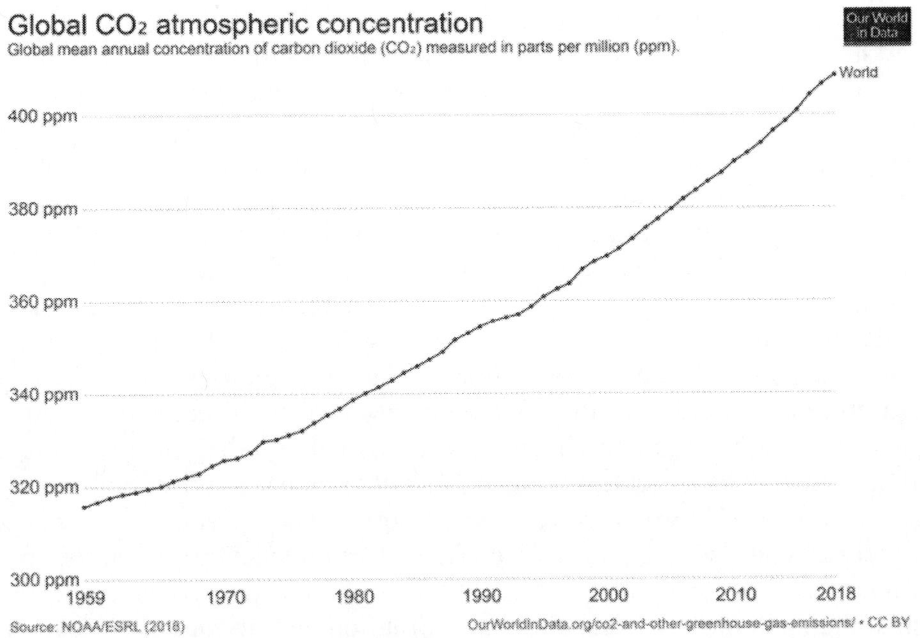

Figure 2. The change of CO_2 concentration in atmosphere (Ritchie and Roser, 2019)

CO_2 can be removed from the atmosphere by the biological carbon cycle, or the photosynthesis, a process by which green plants and some organisms use sunlight to synthesize foods from CO_2 and water. Unfortunately, the amount of CO_2 can be consumed by photosynthesis is lagged behind the CO_2 emission as global population increases. To reduce the concentration of CO_2 in atmosphere and alleviate the global warming, two approaches are employed. One is to reduce the emission of CO_2 through improving fuel efficiency, moving to clean energy or renewable energy such as wind, solar, replacing coal/oil with less carbon-intense fuel (natural gas). Another is the sequestration of CO_2. It is a long-term storage of CO_2 to slow the atmosphere and marine accumulation of CO_2 to mitigate global warming. In the sequestration, CO_2 is captured through biological, chemical, and physical process and then transported and stored in subsurface aquifers, coal beds, depleted oil and gas reservoirs, and ocean water. The storage of CO_2 underground in aquifers, coal beds, depleted oil and gas reservoirs is often referred to geological sequestration.

When selecting a site for geological sequestration, several parameters are considered: reservoir porosity, permeability, existing of fault, sealability of caprock, geometry of reservoir, formation depth, and formation thickness. Porosity and thickness affect the formation storage capacity. Permeability dictates the injection rate and migration of CO_2 in formation. Faults cause the non-uniform flow or tonging phenomenon. Faults in the reservoir are the preferential path for CO_2 to flow. Caprock keeps CO_2 remaining in formation for a long geological time after the injection. Once injected, the CO_2 plume will rise via buoyant forces, since it is less dense than oil and water. Once it encounters a caprock, it will spread laterally until it encounters a gap. If there are fault planes in caprock, there is a possibility the CO_2 could migrate along the fault to the surface. Reservoir geometry controls the distribution and migration direction of CO_2. Formation depth is related to pore pressure, thus impacts the storage capacity and power to inject CO_2.

Numerous researches have been focusing on the storage site selection, dynamics of CO_2 injection and storage in formation, CO_2 migration in reservoir, and possible CO_2 leak through caprock. House et al.

(2003) simulated CO_2 sequestration in two gas fields in North Sea. The gas reservoirs contain 4-9% CO_2 are high permeability sandstone. The authors created two models to estimate the ultimate storage capacity of the reservoir and assess the risk of reservoir leakage. One model assumed a homogeneous sand with properties of the Utsira sand in the Sleipner area. Another based on a hypothetical faulted anticline with high vertical permeability to fault. The results indicated that geologic complexity is a factor influencing migration of CO_2 in the Miocene reservoir sand and the high vertical permeability faults provide high permeability conduits for the CO_2 leak out of the sequestration reservoir. Akanni et al. (2004) developed material balance model for CO_2 sequestration in depleted gas reservoirs. The storage of CO_2 in dry gas, wet gas, and gas condensate reservoir is compared with depleted oil reservoir. The models calculated higher CO_2 storage volume in gas reservoir than oil reservoir and resulted in higher gas recovery than oil recovery. The developed material balance equations can be utilized to predict CO_2 sequestration volume and survey sequestration operation during injection. p/z plot to forecast the sequestration were also presented. Izgec et al. (2005) used experiment and numerical simulation to investigate the injection of CO_2 in deep saline aquifer. The target formation is Midyat carbonate aquifer located in south-east Turkey. Computerized tomography (CT) was used to monitor porosity and permeability change at different CO_2 injection rates, pressures, and temperatures. The effect of salt concentration on the rock property change was also investigated. The experiment was used to calibrate numerical model created by a finite difference, non-isothermal compositional simulator. Solution and dissolution of carbonates via chemical reaction were considered. Bennion and Bachu (2005) presented the relative permeability of three sandstone and three carbonate formation in the Wabamun Lake area southwest of Edmonton in Alberta, Canada, where four coal-fired power plants which produce large volume of CO_2 are located. The formations are representative of the in-situ temperature, pressure, salinity, porosity and permeability characteristics of deep saline aquifers in on-shore North American sedimentary basins. The impact of CO_2-water relative permeability on the CO_2 displacing water performance during the CO_2 injection and on ultimate storage capacity was studied. Spiteri et al. (2005) investigated the hysteresis in the relative permeability of the hydrocarbon phase in a two-phase system. The hysteresis is then evaluated to model geological CO_2 sequestration in saline aquifers. In the sequestration, CO_2 is the non-wetting phase and the trapping of the CO_2 is an essential mechanism during lateral and upward migration of CO_2 plume. Sakurai et al. (2006) described the monitoring of saturation change in a CO_2 sequestration in a pilot test in South Lib erty field, Dayton, Texas. The monitoring was through an observation well that was 100 ft away from the injection well. The depth of the target aquifer is 5000 ft with porosity of 32-35% and air permeability of 2500 mD. 1600 metric tons of CO_2 was injected for eleven days. Because of high formation water salinity, along with high porosity, a pulsed neutron tool was selected as the primary log for monitoring saturation changes through change of thermal neutron absorption cross-section. The C/O ratio measurement and dipole acoustic tool were also used to estimate saturation changes. Breakthrough of CO_2 was observed on the third day. At the end of the injection phase, CO_2 saturation of up to 80% was accounted in the observation well. Even four and a half months after the injection experiments, CO_2 saturation of up to 40% was found in the vicinity of the observation and the injection wells, along with temperature anomalies. Schembre-McCabe et al. (2007) researched the mechanistic of CO_2 sequestration in aquifer and depleted hydrocarbon reservoirs using a compositional simulator. The effects of relative permeability and mobility of different phase, damage due to geochemical reaction, stimulation, and nearby injectors were probed. Their study showed that due to limited storage capacity of a reservoir, injectivity declines during the injection stage. This has impact on injector number. Interference among injectors also affects injector performance. Riaz and Tchelepi (2008) analyzed major factors that control the distribution of CO_2 plume by considering the buoyancy and ignoring trapping, dissolution and chemical reaction. One-dimensional model was applied to characterize a pair of shocks moving in opposite direction: top end and bottom end of the plume. Two-dimensional numerical model was used to understand the plume evolution in terms of viscosity ratio and the capillary number. The study found that the buoyancy force causes the plume to rise up about 500 meter in 700 years. Juanes and MacMinn (2008) proposed a one-dimensional sharp-interface mathematical model for CO_2 migration in saline aquifers. The model accounted for the capillary trapping and gravity override. The CO_2 plume shape was different from previous researches in that the flow of regional groundwater was taken into account. The buoyancy forces lead to gravity override while capillary trapping delays the development of override. Three dimensionless groups: the mobility ratio between CO_2 and water, the gravity number, and trapping coefficient are the key factors that influent shape of CO_2 plume. The capillary trapping coefficient was upscaled from laboratory scale to the basin scale through a formulation based on analytical solution to the model. Nghiem et al. (2009) examined the two most important trapping mechanism for CO_2 storage in saline aquifers: residual gas trapping and solubility trapping. Their study compared how a water injector that is above a CO_2 injector affects the storage of CO_2. In other words, the formation that is used to store CO_2 lies below the saline aquifer. The results indicated that the low permeability aquifer above the CO_2-storage formation can increase the storage capacity while high permeability one does not increase CO_2 trapping. Oloruntobi and LaForce (2009) conducted experimental study to investigate the effect of heterogeneity on buoyancy-driven flow in CO_2 storage. Air was used in the study. The study focused on post-injection period when the migration of CO_2 plume in an aquifer is mainly due to the buoyancy drive that results from the density contrast between the injected CO_2 and formation brine. The study showed that air migration through a homogeneous sand pack provided a more uniform front than the heterogeneous sand pack. A high permeability streak in the sand pack reduced the trapping capacity of the entire pack. The presence of a low-permeability layer in the sand pack increased its trapping capacity. The results also depicted that better consolidated sands trapped more air than poorly consolidated sands. Mantilla et al. (2009) used injection data to assess the migration of plume through adapting a probabilistic history matching software originally developed for oil field applications. The software assimilates injection data commonly monitored at active and inactive wells into models for the subsurface aquifer/reservoir. The algorithm yields an ensemble of realizations geologically consistent with the initial model such that the uncertainty in CO_2 plume location can be easily assessed. The updated models obtained after the history matching process detected the presence of the streak, and the subsequent estimation of CO_2 plume location was much more accurate. Murray et al. (2010) summarized a CO_2 sequestration monitoring in a low formation water salinity reservoir in Japan. The targeted reservoir's water salinity is not sufficiently saline so that uncertainty in the Sigma logging approach increases. To overcome this apparent complication, Sigma logging and alternative measurements such as inelastic ratios and neutron porosity were acquired and combined with the open-hole resistivity, neutron-density and magnetic resonance logs to derive a robust interpretation. Liner et al. (2011) used simulation model to evaluate CO_2 sequestration potential in a deep saline aquifer system in Ness County, Kansas. The study showed that a careful simulation study can maximize CO_2 injection rate, minimize existence of free CO_2, and significantly reduce uncertainty in the safety of CO_2 permanent storage. Amirlatifi et al. (2012) studied the storage of CO_2 in a generic anticline structure. The influence of layer thickness, wavelength and amplitudes at different depths and under different boundary conditions on the maximum CO_2 storage amount was investigated. Their 3-dimensional model was created by finite element analysis preprocessor. Fluid flow under different geometrical and physical conditions was simulated. Pilisi et al. (2012) investigated the technical feasibility of CO_2 sequestration in deepwater offshore Japan. Sub-seabed formations at water depth of 9000 ft in three sites are selected to conduct reservoir simulation. The study indicated that CO_2 capturing technologies and transportation means are at mature stage and drilling vessels are capable to drill well in such deep water. From the technical point of view, CO_2 sequestration in sub-seabed geological formation is feasible. The three studied sites are appropriate for CO_2 storage. Adebayo et al. (2014) emphasized the importance of CO_2 saturation in aquifer during the sequestration. They determined Archie's saturation exponent for CO_2 sequestration in carbonate reservoirs. They proposed a method that can capture the effect of CO_2/brine/rock interaction. The method was compared with conventional porous plate method and it was found that the saturation index differs for both methods largely due to change in pore geometry during CO_2 storage. Park et al. (2015) analyzed the influence of heterogeneous capillary pressure on CO_2 sequestration under different wetting conditions. In their study, numerical simulations are used to estimate the quantity of CO_2 trapped by dissolution, residual, and capillary trapping. Sensitivity analysis is implemented regarding heterogeneity and CO_2/brine wettability. The research depicted that CO_2 leakage decreases as the permeability variation increases. In strongly water-wet system, the largest portion of stored CO_2 is immobilized. Ajibola et al. (2016) believed the dominant mixing mechanism is convective mixing rather than pure diffusion is important as this controls the timescale over which the carbon dioxide-saturated brine mixes with the unsaturated brine. They used a finite difference reservoir simulator to evaluate the predictions of analytical solutions for stability analysis and growth rate of the fingers of different wavenumbers at different Rayleigh numbers. The effects of density difference, permeability anisotropy and diffusion on fingering behavior were investigated through the dimensionless Rayleigh number. The density difference and the vertical permeability were found to mainly control the degree of instability. Ratnakar et al. (2018) predicted gas solubility in brine solutions for CO_2 capture and sequestration. They extended the Setschenow approach by expressing the solubility in terms of ionic strengths and molar concentrations of each salt. The method also characterized each component (gas, anions and cations) against the experimental measurements. A simple methodology was proposed to predict the impact of types of salts on solubility of CO_2.

In the CO_2 sequestration process, the injected CO_2 is displacing water from the injection point and is expected to remain in the reservoir. Due to the nature of one phase displacing another phase in porous media, it is noted that different water saturation exists in the CO_2 plume during the displacement. Water distribution in the plume will affect the size of the plume subsurface. Furthermore, the gravitational segregation between CO_2 and water will cause overriding- tonguing during the injection and impact the shape of plume. To better understand the CO_2 movement underground and development of CO_2 plume, it is necessary to take the two-phase flow and gravity force effects into account when evaluating CO_2 displacing water. The purpose of this study is to develop a simple and practical model to predict the evolution of the shape of CO_2 plume in the injection and post-injection period.

Model Development

The displacement of water by injecting CO_2 is not a piston-like process in aquifer. Because water is the wetting phase and CO_2 is the non-wetting phase when two phases flow in reservoir, water occupies the surface of matrix and small pores while CO2 resides in large pores and centers of pores. As CO_2 is injected into aquifer, CO_2 displaces water away from the injector through the fractional flow. At the same time, CO_2 migrates upward due to the buoyancy forces. As a result, in the horizontal direction, CO_2 migration is controlled by relative permeability and capillary pressure. Meanwhile, in the vertical direction, the upward migration of CO_2 is affected by relative permeability, capillary pressure and buoyancy force.

Figure 3 shows a circular aquifer with an injector located in the center. For CO_2 sequestration in aquifer, Figure 3 can represent the sequestration procedure well. The fractional flow can be viewed as CO_2 displacing water away from the injector. Figure 4 illustrates the flow line and pressure distribution in the reservoir. To make the analysis simple, following assumptions are made:

1) A circular aquifer with constant height
2) Reservoir is homogenous in all rock properties
3) The dip angle of formation is zero

4) CO_2 and water two-phase flow in reservoir
5) Compressibility of water is negligible
6) The variation in water densities can be neglected
7) Reservoir temperature is constant
8) All rock properties do not change during the sequestration.
9) Constant water viscosities during the displacement
10) Average values are used for CO_2 properties such as viscosity, z-factor, and density when CO_2 flows in reservoir
11) Chemical reaction or the dissolution/precipitation can be ignored.

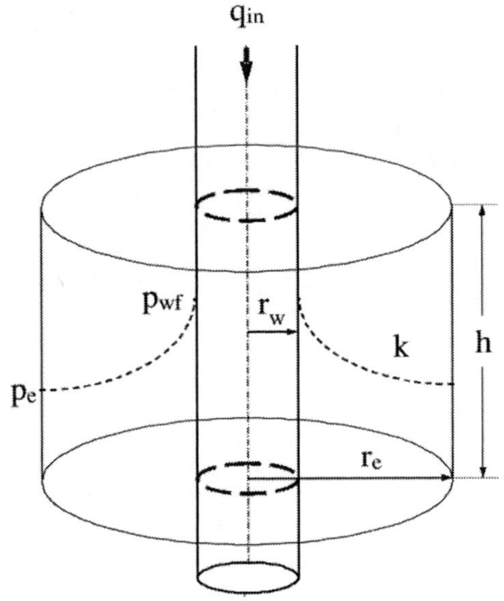

Figure 3. A circular aquifer with an injector located in the center

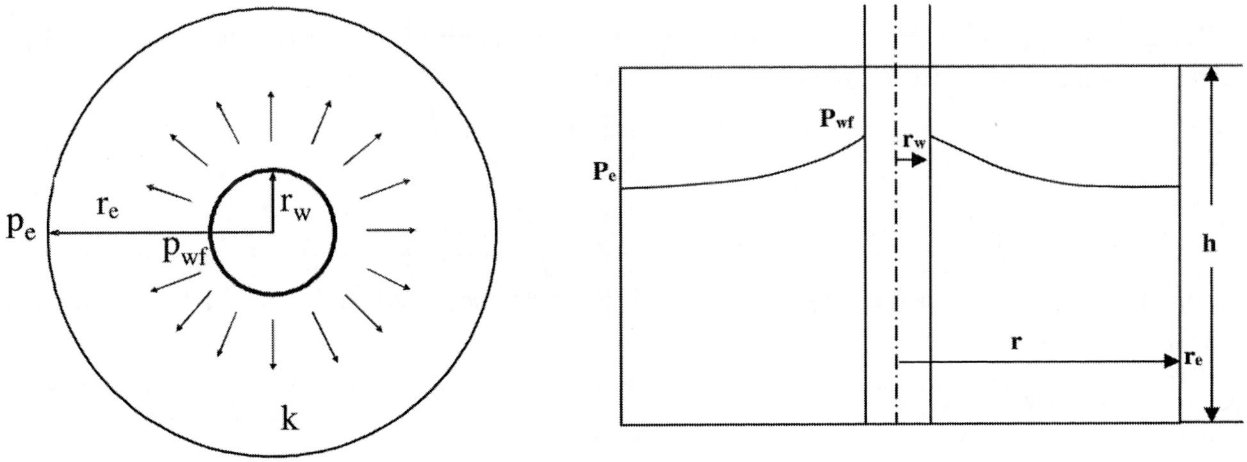

Figure 4. Reservoir system for CO_2 displacing water away from injector: (left) plan view, (right) lateral view

The flow occurs in two directions: horizontal and vertical flows. We first derive the two-phase flow in horizontal direction. Then vertical flow, or segregation of CO_2 and water in the vertical direction, will be discussed next. In the horizontal direction, CO_2 is displacing water away from the injector. Darcy´s equation gives CO_2 (noted as gas in the derivation) and water flow rates

$$q_g = \frac{kk_{rg}}{\mu_g}\frac{\partial(Ap_g)}{\partial r} \quad (1)$$

$$q_w = \frac{kk_{rw}}{\mu_w}\frac{\partial(Ap_w)}{\partial r} \quad (2)$$

where

A= flow area

k = reservoir permeability

k_{rg} = relative permeability to gas

k_{rw} = relative permeability to water

p_g = gas pressure

p_w = water pressure

q_g = gas rate

q_w = water rate

r = radius from wellbore

μ_g = gas viscosity

μ_w = water viscosity

In the displacement, CO_2 is the non-wetting phase and water is the wetting phase. Recalling the concept of capillary pressure, we have

$$P_c = p_g - p_w \quad (3)$$

where

P_c = capillary pressure

Replacing water pressure by gas and capillary pressure **Equation** (2) becomes

$$q_w = \frac{kk_{rw}}{\mu_w}\frac{\partial[A(p_g-p_c)]}{\partial r} \quad (4)$$

Expressing in pressure gradient, **Equations** (1) and (4) are changed to

$$\frac{\partial(Ap_g)}{\partial r} = \frac{\mu_g}{kk_{rg}}q_g \quad (5)$$

$$\frac{\partial(Ap_g)}{\partial r} - \frac{\partial(AP_c)}{\partial r} = \frac{\mu_w}{kk_{rw}}q_w \quad (6)$$

Subtracting **Equation** (5) from (6) we obtain

$$-\frac{\partial(AP_c)}{\partial r} = \frac{\mu_w}{kk_{rw}}q_w - \frac{\mu_g}{kk_{rg}}q_g$$

or

$$\frac{\partial(AP_c)}{\partial r} = \frac{1}{k}\left(\frac{\mu_g}{k_{rg}}q_g - \frac{\mu_w}{k_{rw}}q_w\right) \quad (7)$$

At this stage we can introduce the concepts of total fluid rate and fractional flow, which are defined as

$$q_t = q_g + q_w \quad (8)$$

$$f_g = \frac{q_g}{q_t} \text{ and } f_w = \frac{q_w}{q_t} \quad (9)$$

where

q_t = total fluid rate

f_g = gas fraction

f_w = water fraction

Substituting **Equations** (8) and (9) into (7) yields

$$f_g = \frac{1+\frac{\partial(AP_c)}{\partial r}\frac{kk_{rw}}{q_t\mu_w}}{1+\frac{k_{rw}\mu_g}{k_{rg}\mu_w}} \quad (10)$$

Flow area is defined as

$$A = 2\pi rh \quad (11)$$

where

h = reservoir thickness

Substituting **Equation** (11) into (10) we have

$$f_g = \frac{1+\frac{2\pi hkk_{rw}}{q_t\mu_w}\left(\frac{r\partial P_c}{\partial r}+P_c\right)}{1+\frac{k_{rw}\mu_g}{k_{rg}\mu_w}} \quad (12)$$

where

$$\frac{\partial P_c}{\partial r} = \frac{\partial P_c}{\partial S_g}\frac{\partial S_g}{\partial r} \quad (13)$$

It should be noted that capillary pressure decreases as radius increases in this case. Therefore $\frac{\partial P_c}{\partial r}$ is negative. If capillary pressure is negligible, **Equation** (12) is simplified to

$$f_g = \frac{1}{1+\frac{k_{rw}\mu_g}{k_{rg}\mu_w}} \quad (14)$$

According to Brooks-Corey (1954 and 1964) capillary pressure model, drainage capillary pressure in the CO_2 sequestration in aquifer can be expressed as

$$P_c = P_d\left(\frac{S_w-S_{wirr}}{1-S_{wirr}}\right)^{\frac{-1}{\lambda}} \quad (15)$$

where

S_{wirr} = the irreducible water saturation

P_d = the threshold pressure.

Brooks and Corey (1954 and 1964) related the parameter λ to the distribution of pore sizes. For narrow distributions, λ is greater than 2; for wide distributions, λ is less than 2. **Equation** (12) indicates that gas fraction is a function of location and injection volume (or injection time if injection rate is kept constant). Now we derive the continuity equation of radial fractional flow. Considering the CO_2 displacing water, material balance equation provides that the mass change in a control volume for a time period can be shown as **Figure** 5. Since the flow direction is from injector to aquifer outer boundary, we define the center of wellbore as the start point, where r=0, and aquifer outer boundary as the end point where $r=r_e$. Therefore, in the dimensionless analysis, dimensionless radius can be defined as, $r_D=r/r_e$, the start point has r_D=0, and the end point has r_D=1. Material balance gives

$$\left[(q_g\rho_g)_r - (q_g\rho_g)_{r+\Delta r}\right]\Delta t = \pi h[(r+\Delta r)^2 - r^2]\varphi\left[(S_g\rho_g)^{t+\Delta t} - (S_g\rho_g)^t\right] \quad (16)$$

where

S_g = gas saturation

ρ_g = gas density

Δt = time period

t = time

Δr= radius incremental

r= radius

r_e= distance between reservoir outer boundary to wellbore

φ = porosity

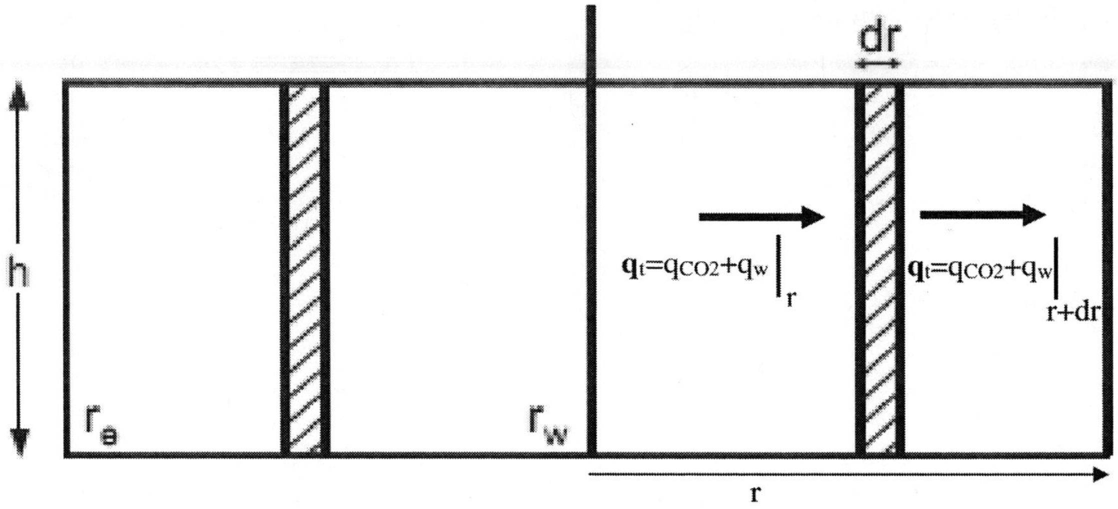

Figure 5. A control volume in a circular aquifer with an injector located in the center.

Simplifying **Equation** (16) we have

$$\left[(q_g\rho_g)_r - (q_g\rho_g)_{r+\Delta r}\right]\Delta t = \pi h[2r\Delta r + (\Delta r)^2]\varphi\left[(S_g\rho_g)^{t+\Delta t} - (S_g\rho_g)^{t}\right] \quad (17)$$

As $\Delta r \to 0$ and $\Delta t \to 0$, we have

$$2r\Delta r + (\Delta r)^2 \approx 2r\Delta r \quad (18)$$

Equation (17) becomes partial differential equation

$$-\frac{\partial(q_g\rho_g)}{\partial r} = 2r\pi h\varphi\frac{\partial(S_g\rho_g)}{\partial t} \quad (19)$$

If an average gas density is used and is assumed constant, we have

$$-\frac{\partial q_g}{\partial r} = 2r\pi h\varphi\frac{\partial S_g}{\partial t} \quad (20)$$

Substituting **Equation** (9) into (20) gives

$$-\frac{\partial(f_g q_t)}{\partial r} = 2r\pi h\varphi\frac{\partial S_g}{\partial t} \quad (21)$$

If CO_2 injection rate is constant and an average CO_2 density at aquifer temperature and average pressure (average of injection pressure and aquifer outer boundary pressure) is used to represent CO_2 density in aquifer, we have a constant total fluid rate. **Equation** (21) can be simplified to

$$-\frac{\partial f_g}{\partial r} = \frac{2r\pi h\varphi}{q_t}\frac{\partial S_g}{\partial t} \quad (22)$$

Since water fraction is function of water saturation, $f_g(S_g)$, applying chain rule to partial differential equation results in

$$-\frac{df_g}{dS_g}\frac{\partial S_g}{\partial r} = \frac{2r\pi h\varphi}{q_t}\frac{\partial S_g}{\partial t} \quad (23)$$

Observing that water saturation is function of time, t, and position, r, we can express

$$dS_g = \frac{\partial S_g}{\partial t}dt + \frac{\partial S_g}{\partial r}dr \quad (24)$$

The fact that at the displacement front the CO_2 saturation is constant provides us a boundary condition.

$$dS_g = \frac{\partial S_g}{\partial t}dt + \frac{\partial S_g}{\partial r}dr = 0$$

or

$$\frac{\partial S_g}{\partial r} = -\frac{\partial S_g}{\partial t}\frac{dt}{dr} \quad (25)$$

Substituting **Equation** (25) into (23) yields

$$\frac{df_g}{dS_g}dt = \frac{2r\pi h\varphi}{q_t}dr \quad (26)$$

Integrating **Equation** (26) yields an equation for displacement front position, r_f.

$$r_f^2 = \frac{tq_t}{\pi h\varphi}\left(\frac{df_g}{dS_g}\right)_f \quad (27)$$

where

r_f = displacement front position in radial aquifer

There are two solutions to **Equation** (27), which are

$$r_f = \pm\sqrt{\frac{tq_t}{\pi h\varphi}\left(\frac{df_g}{dS_g}\right)_f} \quad (28)$$

Obviously only one solution is correct to match with the physical phenomenon. Considering the fact that location can only be positive, therefore we can eliminate the solution

$$r_f = -\sqrt{\frac{tq_t}{\pi h\varphi}\left(\frac{df_g}{dS_g}\right)_f} \quad (29)$$

Therefore, the correct solution is

$$r_f = \sqrt{\frac{tq_t}{\pi h\varphi}\left(\frac{df_g}{dS_g}\right)_f} \quad (30)$$

For any CO_2 saturation that is behind the displacement front, S_g, the position can be calculated by

$$r_{Sg} = \sqrt{\frac{tq_t}{\pi h\varphi}\left(\frac{df_g}{dS_g}\right)_{Sg}} \quad (31)$$

where

r_{Sg} = position of any CO_2 saturation that is behind the displacement front

Water saturation ahead of the displacement front is the initial water saturation in reservoir. For aquifer, initial water saturation can be assumed 100%. CO_2 saturation ahead of displacement front is 0. Therefore, locations of different CO_2 saturations can be estimated accordingly based on **Equations** (31). **Figure** 7 shows the plot of CO_2 saturation versus radius at different injection times assuming vertical segregation of CO_2 and water can be ignored.

Because of the density difference between CO_2 and water. Segregation between CO_2 and water occurs when CO_2 mixes with water in reservoir. In the segregation, water flows downward while CO_2 flows upward. Considering a control volume around the well (or a ring-shape of reservoir volume) at a radius of r, material balance gives

$$[(q_{wV}\rho_w)_{z+\Delta z} - (q_{wV}\rho_w)_z]\Delta t = A\Delta z\varphi[(S_w\rho_w)^{t+\Delta t} - (S_w\rho_w)^t] \quad (32)$$

where

q_{wV} = water flowrate in vertical direction

ρ_w = water density

A = flow area of control volume when vertical flow occurs

Δz= incremental in vertical direction

S_w = water saturation

As $\Delta z \to 0$ and $\Delta t \to 0$, we have

Equation (32) becomes partial differential equation

$$\frac{\partial(q_{wV}\rho_w)}{\partial z} = A\varphi\frac{\partial(S_w\rho_w)}{\partial t} \quad (33)$$

Assuming constant water density, **Equation (33)** can be simplified to

$$\frac{\partial(q_{wV})}{\partial z} = A\varphi\frac{\partial(S_w)}{\partial t} \qquad (34)$$

The flow of fluid in vertical direction is controlled by buoyancy forces, capillary pressure, and vertical permeability. While the buoyancy force facilitates the water downward migration, the effect of capillary pressure depends on the wettability. In CO_2-water system, water is wetting phase and CO_2 is non-wetting phase. Therefore, capillary pressure facilitates water flow as well. Darcy law gives

$$q_{wV} = A\frac{k_V k_{rw}}{\mu_w}\frac{\Delta z(\rho_w-\rho_g)+P_c}{\Delta z} \qquad (35)$$

where

k_V = vertical permeability

Substituting **Equation (35)** into **(34)** yields

$$\frac{\frac{k_V}{\mu_w}(\rho_w-\rho_g)\partial(k_{rw})}{\partial z} + \frac{\partial\left(\frac{P_c}{\Delta z}\right)}{\partial z} = \varphi\frac{\partial(S_w)}{\partial t} \qquad (36)$$

If capillary pressure is negligible, **Equation (36)** can be simplified to

$$\frac{\frac{k_V}{\mu_w}(\rho_w-\rho_g)\partial(k_{rw})}{\partial z} = \varphi\frac{\partial(S_w)}{\partial t} \qquad (37)$$

Material balance is also applied to CO_2 vertical flow

$$\left[(q_{gV}\rho_g)_z - (q_{gV}\rho_g)_{z+\Delta z}\right]\Delta t = A\Delta z\varphi\left[(S_g\rho_g)^{t+\Delta t} - (S_g\rho_g)^{t}\right] \qquad (38)$$

where

q_{gV} = gas flowrate in vertical direction

As $\Delta z \to 0$ and $\Delta t \to 0$, we have

Equation (38) becomes partial differential equation

$$-\frac{\partial(q_{gV}\rho_g)}{\partial z} = A\varphi\frac{\partial(S_g\rho_g)}{\partial t} \qquad (39)$$

Assuming average gas density is used, **Equation (39)** can be simplified to

$$-\frac{\partial(q_{gV})}{\partial z} = A\varphi\frac{\partial(S_g)}{\partial t} \qquad (40)$$

Buoyancy force facilitates the gas upward migration while capillary pressure has the opposite effect. Darcy law gives

$$q_{gV} = A\frac{k_V k_{rg}}{\mu_g}\frac{\Delta z(\rho_w-\rho_g)-P_c}{\Delta z} \qquad (41)$$

Substituting **Equation (41)** into **(40)** yields

$$-\frac{\frac{k_V}{\mu_g}(\rho_w-\rho_g)\partial(k_{rg})}{\partial z} + \frac{\partial\left(\frac{P_c}{\Delta z}\right)}{\partial z} = \varphi\frac{\partial(S_g)}{\partial t} \qquad (42)$$

If capillary pressure is negligible, **Equation (42)** can be simplified to

$$-\frac{\frac{k_V}{\mu_g}(\rho_w-\rho_g)\partial(k_{rg})}{\partial z} = \varphi\frac{\partial(S_g)}{\partial t} \qquad (43)$$

Coupling **Equation** (31) for horizontal flow with **Equations (37)** and **(43)** for vertical CO_2-water segregation, CO_2 saturation distribution in the aquifer at different injection times or post-injection periods can be calculated.

Case Study

A case study was conducted to illustrate the analysis of CO_2 displacing water in aquifer. The input data are shown in **Tables** 3. **Table** 4 is the relative permeabilities versus gas and water saturations. Figure 6 shows the capillary pressures of CO_2-water system versus water saturation used in the case study. It is a drainage curve that simulates non-wetting phase (CO_2) displacing water (wetting phase).

Table 3. The input data for CO_2 displacing water in an aquifer system

CO_2 injection rate at standard conditions	=	20,000	Mscf/day
Aquifer radius	=	10,000	ft
Aquifer thickness	=	100	ft
Initial water saturation	=	100%	
Irreducible water saturation	=	20.0%	
Residual CO_2 saturation	=	25.0%	
CO_2 viscosity	=	0.02	cp
Water viscosity	=	1.0	cp
Porosity	=	20.0%	
Reservoir horizontal permeability	=	200	mD
Reservoir vertical permeability	=	20, 100, or 200	mD
Reservoir temperature	=	150	°F
Average pressure in injection calculation	=	3500	psia
Aquifer porosity	=	0.25	
CO_2 compressibility factor at reservoir conditions	=	1	

Table 4. The relative permeabilities versus water saturation

No.	Sw	Sg	krw	krg
	0	1	0	0.5
1	0.2	0.8	0	0.5
2	0.209	0.791	0.0001	0.4838
3	0.218	0.782	0.0005	0.4678
4	0.227	0.773	0.0012	0.4522
5	0.236	0.764	0.0021	0.4368
6	0.245	0.755	0.0033	0.4216
7	0.254	0.746	0.0048	0.4067
8	0.263	0.737	0.0065	0.3921
9	0.272	0.728	0.0085	0.3778
10	0.281	0.719	0.0108	0.3637
11	0.290	0.710	0.0134	0.3499
12	0.299	0.701	0.0162	0.3364
13	0.308	0.692	0.0192	0.3231
14	0.317	0.683	0.0226	0.3101
15	0.326	0.674	0.0262	0.2974
16	0.335	0.665	0.0300	0.2849
17	0.344	0.656	0.0342	0.2727
18	0.353	0.647	0.0386	0.2608

19	0.362	0.638	0.0433	0.2491
20	0.371	0.629	0.0482	0.2377
21	0.380	0.620	0.0534	0.2266
22	0.389	0.611	0.0589	0.2157
23	0.398	0.602	0.0646	0.2051
24	0.407	0.593	0.0706	0.1948
25	0.416	0.584	0.0769	0.1847
26	0.425	0.575	0.0835	0.1749
27	0.434	0.566	0.0903	0.1654
28	0.443	0.557	0.0974	0.1561
29	0.452	0.548	0.1047	0.1471
30	0.461	0.539	0.1123	0.1384
31	0.470	0.530	0.1202	0.1299
32	0.479	0.521	0.1283	0.1217
33	0.488	0.512	0.1368	0.1138
34	0.497	0.503	0.1454	0.1061
35	0.506	0.494	0.1544	0.0987
36	0.515	0.485	0.1636	0.0916
37	0.524	0.476	0.1731	0.0847
38	0.533	0.467	0.1828	0.0781
39	0.542	0.458	0.1928	0.0718
40	0.551	0.449	0.2031	0.0657
41	0.560	0.440	0.2137	0.0600
42	0.569	0.431	0.2245	0.0544
43	0.578	0.422	0.2356	0.0492
44	0.587	0.413	0.2469	0.0442
45	0.596	0.404	0.2586	0.0395
46	0.604	0.396	0.2704	0.0350
47	0.613	0.387	0.2826	0.0308
48	0.622	0.378	0.2950	0.0269
49	0.631	0.369	0.3077	0.0232
50	0.640	0.360	0.3207	0.0198
51	0.649	0.351	0.3339	0.0167
52	0.658	0.342	0.3474	0.0139
53	0.667	0.333	0.3611	0.0113
54	0.676	0.324	0.3751	0.0090
55	0.685	0.315	0.3894	0.0069
56	0.694	0.306	0.4040	0.0051
57	0.703	0.297	0.4188	0.0036
58	0.712	0.288	0.4339	0.0023
59	0.721	0.279	0.4493	0.0014
60	0.730	0.270	0.4649	0.0006

61	0.739	0.261	0.4808	0.0002
62	0.748	0.252	0.4969	0.0000
63	0.757	0.243	0.5000	0.0000
64	0.766	0.234	0.5000	0.0000
65	0.775	0.225	0.5000	0.0000
66	0.784	0.216	0.5000	0.0000
67	0.793	0.207	0.5000	0.0000
68	0.802	0.198	0.5000	0.0000
69	0.811	0.189	0.5000	0.0000
70	0.820	0.180	0.5000	0.0000
71	0.829	0.171	0.5000	0.0000
72	0.838	0.162	0.5000	0.0000
73	0.847	0.153	0.5000	0.0000
74	0.856	0.144	0.5000	0.0000
75	0.865	0.135	0.5000	0.0000
76	0.874	0.126	0.5000	0.0000
77	0.883	0.117	0.5000	0.0000
78	0.892	0.108	0.5000	0.0000
79	0.901	0.099	0.5000	0.0000
80	0.910	0.090	0.5000	0.0000
81	0.919	0.081	0.5000	0.0000
82	0.928	0.072	0.5000	0.0000
83	0.937	0.063	0.5000	0.0000
84	0.946	0.054	0.5000	0.0000
85	0.955	0.045	0.5000	0.0000
86	0.964	0.036	0.5000	0.0000
87	0.973	0.027	0.5000	0.0000
88	0.982	0.018	0.5000	0.0000
89	0.991	0.009	0.5000	0.0000
90	1.000	0.000	0.5000	0.0000

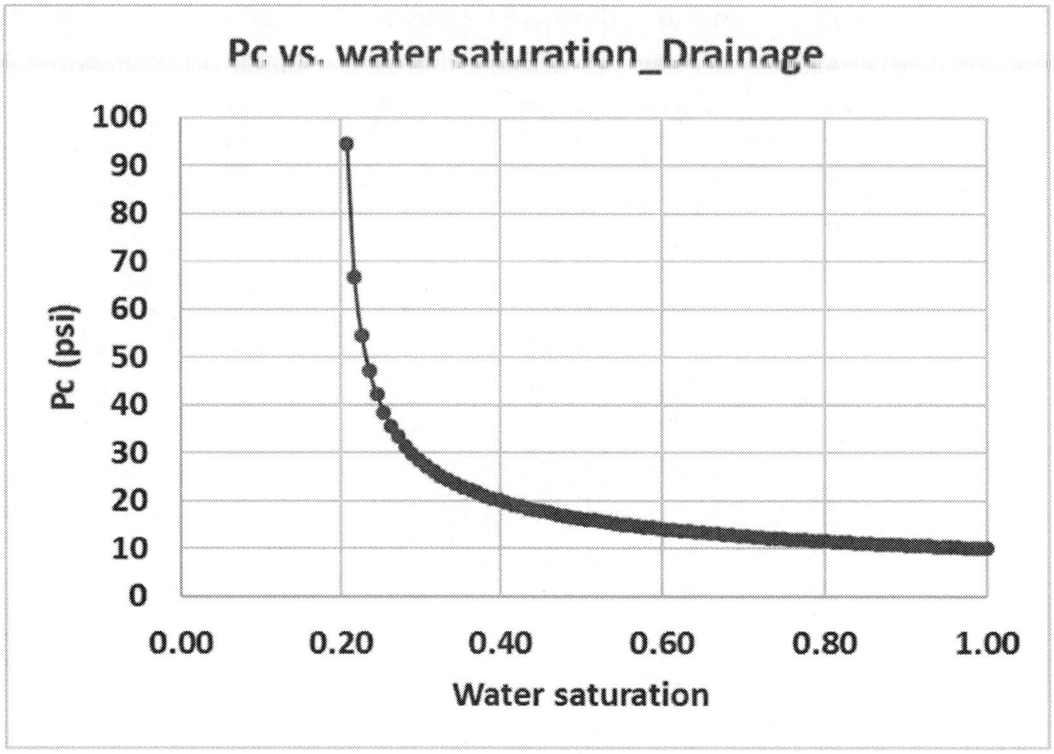

Figure 6. Capillary pressure versus water saturation for the CO_2 displacing water in aquifer during injection

If vertical segregation is ignored, CO_2 saturation at any radius is a unique number and is constant at different aquifer elevations. **Figure 7** illustrates the CO_2 saturation distribution along the radius of aquifer at different injection times. In this case, CO_2 saturation at the displacement front is 0.342. Ahead of the displacement front, water saturation is 100%, or CO_2 saturation is 0.0%. The location of displacement front at different injection time can be found in **Table 5**. **Figure 7** and **Table 5** indicate that movement of CO_2 saturation away from the well is not linear to the injection time. The movement slows down because more pore volumes needed to be filled as radius increases.

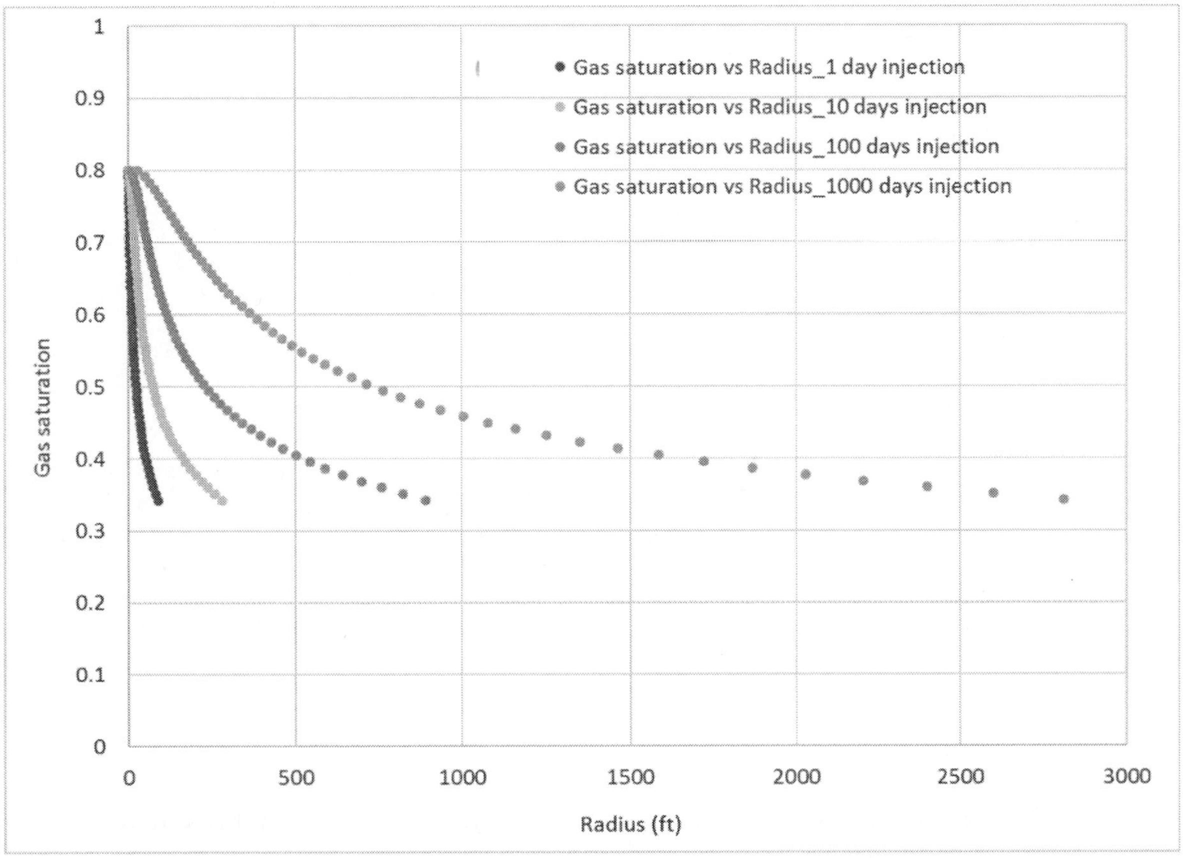

Figure 7. Change of CO_2 saturation in aquifer during different injection times

Table 5. Location of displacement front at different injection times

Injection time (days)	1	2	5	10	20	50	100	200	500	1000	2000	5000	10000
Displacement front Location (ft)	89	126	199	281	398	629	889	1257	1988	2811	3976	6286	8890

CO_2-water segregation causes CO_2 to migrate to the top of aquifer. **Figure 8** depicts the variation of CO_2 saturation from the bottom of aquifer to the top of aquifer at different radii. It can be seen that near the wellbore the segregation is more complete than away from the wellbore, which is indicated by wide ranges of CO_2 saturation between the bottom of aquifer (with aquifer elevation of 0 ft) and the top of aquifer (with aquifer elevation of 100 ft). The segregation leads to the overriding of CO_2 on water during the injection and for post-injection period. Increasing injection rate can reduce the overriding effect temporary, but segregation will occur until the buoyancy force is balanced by the capillary force. At the beginning of the segregation, the difference between the buoyancy force and capillary force is the highest. The force difference that causes the CO_2 vertical-migration is the largest. As gas migration continues, capillary force increases. Therefore, the segregation speed is highest at the beginning and then slows down as time goes on. Once the buoyancy force is equal to capillary force, CO_2-water system reaches a dynamic equilibrium status and CO_2 will remain in location until the equilibrium is broken.

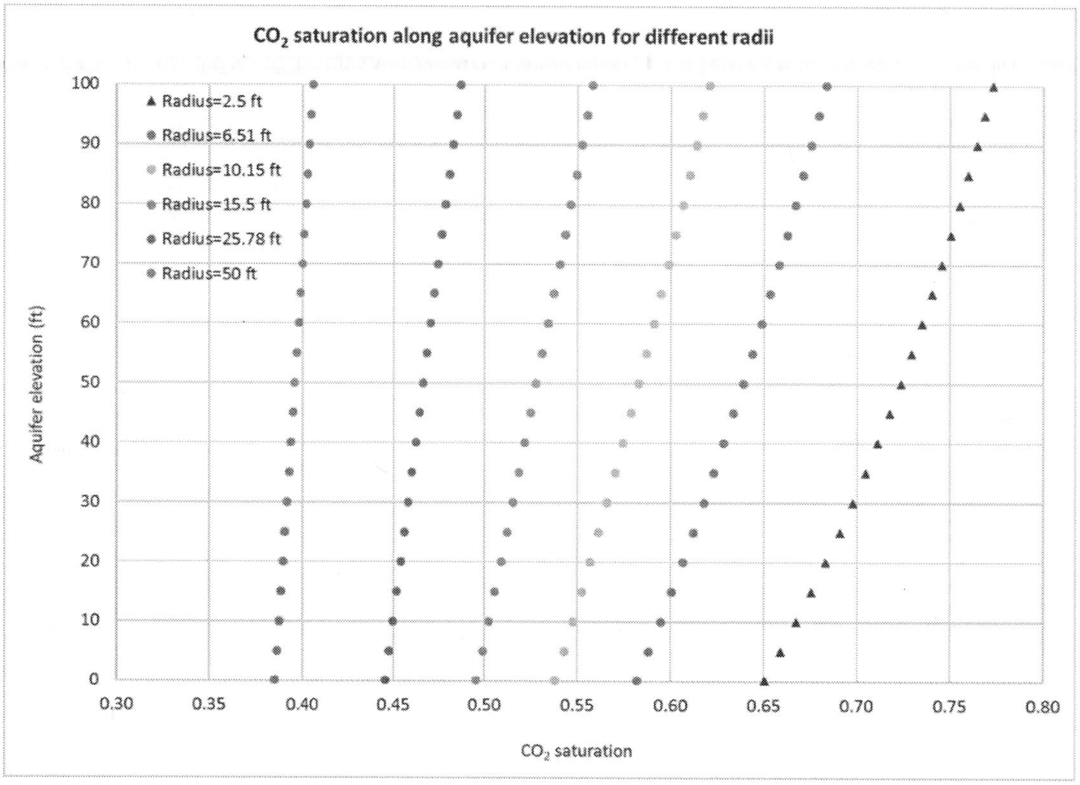

Figure 8. Change of CO_2 saturation along aquifer elevation at different radii

Anisotropy of rock properties is often observed in aquifer. For most aquifers vertical permeability is less than horizontal permeability, which alleviates the CO_2 vertical migration. In this study, the effect of vertical/horizontal permeability ratio on the CO_2 overriding on water is investigated. Three cases with vertical/horizontal permeability ratio of 1.0, 0.5 and 0.1 are compared to analyze how low vertical permeability delays the CO_2 upward migration. One should note that low permeability also means high capillary pressure, which also limits the upward flow of CO_2. **Figure 9** shows the CO_2 saturation distribution along aquifer elevation for different vertical/horizontal permeability ratios at two radii. It clearly shows that low vertical/horizontal permeability ratio has less CO_2 overriding than that of high vertical/horizontal permeability ratio.

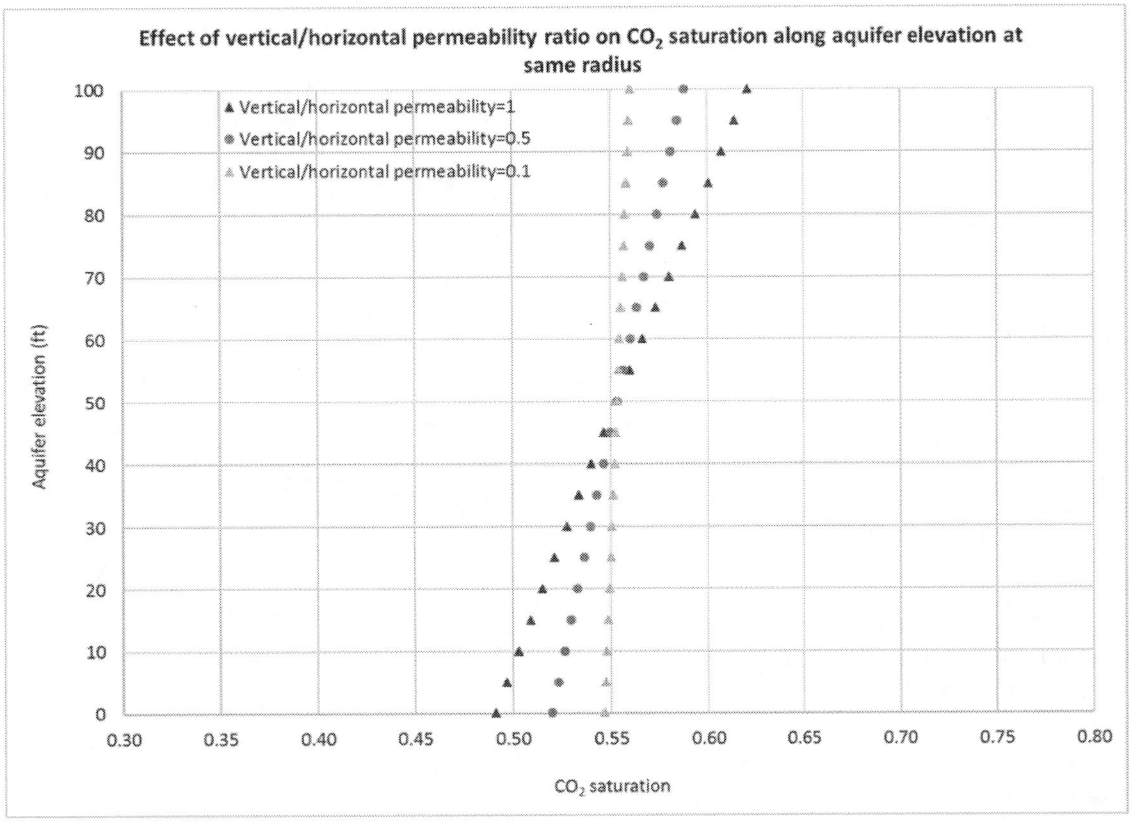

Figure 9. Effect of permeability anisotropy on CO_2-water segregation at same radius

Conclusions

Following conclusions can be drawn upon this study.

- A model coupling CO_2-water fractional flow in horizontal direction with CO2-water segregation in vertical direction has been proposed. It can be used to simulate the performance of CO_2 sequestration in aquifer.
- The model indicates that CO_2 displacing water is not a piston-like displacement. CO_2 saturation at the displacement front is a function of fluid and rock properties. Behind the CO_2 front, CO_2 saturation increases gradually, and water saturation decreases accordingly until water saturation reaches irreducible water saturation.
- The propagation of CO_2 front is not linear to the injection time. Its movement slows down as it moves away from injection well.
- For same radius, CO_2-water segregation causes higher CO_2 saturation at top of aquifer and low at the bottom of aquifer.
- Low vertical/horizontal permeability ratio can delay CO_2-water segregation, leads to a less severity of CO_2 overriding.

Acknowledgement

The authors are grateful to The Petroleum Engineering Department at the University of North Dakota. This research is supported in part by the North Dakota EPSCoR Program under award number EPS-0814442 and by Peng Yu: Guangxi Natural Science Foundation (No.2016GXNSFBA380180 and No. 2017GXNSFAA198105).

Nomenclature

A= flow area
A = flow area of control volume when vertical flow occurs
f_g = gas fraction
f_w = water fraction
h = reservoir thickness
k = reservoir permeability
k_{rg} = relative permeability to gas
k_{rw} = relative permeability to water
k_V = vertical permeability
P_c = capillary pressure
P_d = the threshold pressure.
p_g = gas pressure
p_w = water pressure
q_g = gas rate
q_{gV} = gas flowrate in vertical direction
q_t = total fluid rate
q_w = water rate
q_{wV} = water flowrate in vertical direction
r = radius from wellbore
r_e= distance between reservoir outer boundary to wellbore
r_f= displacement front position in radial aquifer
r_{Sg}= position of any CO_2 saturation that is behind the displacement front
S_g = gas saturation
S_w = water saturation
S_{wirr} = irreducible water saturation
t = time
μ_g = gas viscosity
μ_w = water viscosity
ρ_g = gas density
ρ_w = water density
φ = porosity
Δt = time period
Δr= radius incremental
Δz= incremental in vertical direction

References

1. Adebayo, A. R., Al-Yousef, H. Y., & Mahmoud, M. (2014, August 5), "Determination of Archie's Saturation Exponent for CO_2 Sequestration in Carbonate Reservoirs", *Society of Petroleum Engineers*, doi:10.2118/172354-MS.
2. Ajibola, J., Adam, A., & Muggeridge, A. (2016, October 25), "Gravity Driven Fingering and Mixing During CO_2 Sequestration", *Society of Petroleum Engineers*, doi:10.2118/182317-MS.
3. Amirlatifi, A., Eckert, A., Nygaard, R., Bai, B., Liu, X., & Paradeis, M. (2012, January 1), "Role of Geometrical Influences of CO_2 Sequestration in Anticlines", *American Rock Mechanics Association*.
4. Bennion, B., & Bachu, S. (2005, January 1), "Relative Permeability Characteristics for Supercritical CO_2 Displacing Water in a Variety of Potential Sequestration Zones", *Society of Petroleum Engineers*, doi:10.2118/95547-MS.
5. Brooks, R.H. and Corey, A.T. (1964), "Hydraulic properties of porous media", *Hydrology Paper No. 3*, Colorado State University, Fort Collins, Colorado, 22–27.
6. Corey, A.T. (1954), "The interrelation between gas and oil relative permeabilities", *Producers Monthly* 19 (November): 38–41.
7. Frailey, S. M. (2004, January 1), "Material Balance Reservoir Model for CO_2 Sequestration in Depleted Gas Reservoirs", *Society of Petroleum Engineers*, doi:10.2118/90669-MS.
8. Hannah Ritchie and Max Roser (2019), "CO_2 and other Greenhouse Gas Emissions", Published online at *OurWorldInData.org*, Retrieved from https://ourworldindata.org/co2-and-other-greenhouse-gas-emissions.
9. Haynes, H. M., ed. (2016), *CRC Handbook of Chemistry and Physics*(97th ed.), CRC Press, p. 14-3, ISBN 978-1-4987-5428-6.
10. House, N. J., Faulder, D. D., Olson, G. L., & Fanchi, J. R. (2003, January 1), "Simulation Study of CO_2 Sequestration in a North Sea Formation. Society of Petroleum Engineers", doi:10.2118/81202-MS.
11. Izgec, O., Demiral, B., Bertin, H. J., & Akin, S. (2005, January 1), "Experimental and Numerical Investigation of Carbon Sequestration in Saline Aquifers", *Society of Petroleum Engineers*, doi:10.2118/94697-STU.
12. Juanes, R., & MacMinn, C. (2008, January 1), "Upscaling of Capillary Trapping Under Gravity Override: Application to CO_2 Sequestration in Aquifers", Society of Petroleum Engineers. doi:10.2118/113496-MS.
13. Liner, C., Geng, P., Zeng, J., King, H., & Li, J. (2011, January 1), "A CO_2 Sequestration Simulation Case Study at the Dickman Field", Ness Co., Kansas. Society of Petroleum Engineers, doi:10.2118/145791-MS.
14. Mantilla, C. A., Srinivasan, S., Cross III, E. A., & Bryant, S. L. (2009, January 1), "Inexpensive Assessment of Plume Migration during CO_2 Sequestration", *Society of Petroleum Engineers*, doi:10.2118/126693-MS.
15. Murray, D. R., Yang, X., Horie, T., Yoshimura, T., & Mito, S. (2010, January 1), "CO_2 Sequestration Monitoring in a Low Salinity Reservoir", *Society of Petroleum Engineers*, doi:10.2118/130773-MS.
16. Nghiem, L. X., Yang, C., Shrivastava, V. K., Kohse, B. F., Hassam, M. S., Chen, D., & Card, C. (2009, January 1), "Optimization of Residual Gas and Solubility Trapping for CO_2 Sequestration in Saline Aquifers", *Society of Petroleum Engineers,* doi:10.2118/119080-MS.
17. Oloruntobi, O. S., & LaForce, T. C. (2009, January 1), "Effect of Aquifer Heterogeneity on CO_2 Sequestration", *Society of Petroleum Engineers*, doi:10.2118/121776-MS.

18. Park, S., J., Kim, T. H., & Lee, K. S. (2015, July 27), "Influence of Heterogeneous Capillary Pressure on CO_2 Sequestration in Different Wetting Conditions", *International Society of Offshore and Polar Engineers.*
19. Pilisi, N., Ghorbani, D., & Vasantharajan, S. (2012, January 1), "CO_2 Sequestration in Deepwater Sediments Offshore Japan", *Carbon Management Technology Conference*, doi:10.7122/151756-MS.
20. Ratnakar, R. R., Venkatraman, A., Kalra, A., & Dindoruk, B. (2018, September 24), "On the Prediction of Gas Solubility in Brine Solutions for Applications of CO_2 Capture and Sequestration", *Society of Petroleum Engineers*, doi:10.2118/191541-MS.
21. Riaz, A., & Tchelepi, H. A. (2006, January 1), "Unstable Dynamics of Vertical Displacement in Porous Media Associated With CO_2 Sequestration", *Society of Petroleum Engineers*, doi:10.2118/103169-MS.
22. Sakurai, S., Ramakrishnan, T. S., Boyd, A., Mueller, N., & Hovorka, S. (2006, December 1), "Monitoring Saturation Changes for CO_2 Sequestration: Petrophysical Support of the Frio Brine Pilot Experiment", *Society of Petrophysicists and Well-Log Analysts.*
23. Schembre-McCabe, J. M., Kamath, J., & Gurton, R. M. (2007, January 1), "Mechanistic Studies of CO_2 Sequestration", *International Petroleum Technology Conference*, doi:10.2523/IPTC-11391-MS.
24. Spiteri, E., Juanes, R., Blunt, M. J., & Orr, F. M. (2005, January 1), "Relative-Permeability Hysteresis: Trapping Models and Application to Geological CO_2 Sequestration", *Society of Petroleum Engineers*, doi:10.2118/96448-MS.
25. U.S. EPA, (2019), https://web.archive.org/web/20041112075806/http://www.epa.gov/nonco2/econ-inv/table.html. Retrieved in June 2019.
26. Wallace, John M. and Peter V. Hobbs. (2006), "Atmospheric Science: An Introductory Survey", *Elsevier*, Second Edition, ISBN 978-0-12-732951-2, Chapter 1.

The Development of Robust Particle Gel for CO2 Flooding

Authors:
Long, Y., Missouri University of Science and Technology

As a significant Carbon Capture and Storage (CCS) technology, Carbon dioxide-Enhanced oil recovery (CO2-EOR) contributed to reducing 10-20 billion tons of Greenhouse Gas (GHG) emission in U.S. The rapid increase of CO2-EOR projects indicated its enormous industrial value wherein 127 projects have been established across the country. However, the sweep efficiency of CO2 is still limited owing to its unfavorable mobility and heterogeneity of oil reservoir. Gel treatment is a low-cost, high-efficiency, and widely applicable technology to remedy such conformance problems. In spite of the versatility, conventional gel system based on copolymers of polyacrylate (PAA) and polyacrylamide (PAM) turned out to be a deficient plugging agent for CO2 flooding due to its syneresis in acidic conditions.
To improve gel performance in CO2 condition, an efficient CO2-responsive material, dimethylaminoethyl methacrylate (DMAEMA), was introduced to the gel system. In our work, particle gels containing CO2-responsive moieties, amidine group (-NCH3CH3), were successfully synthesized and optimized. Herein, a pressure-tolerate acrylic cylinder was used to assess gel swelling capacity in which CO2 presented. Moreover, the swelling performance was examined in different buffers pH varied from 2 to 7. Finally, a comprehensive characterization regarding gel rheology was carried out. Consequently, CO2-responsive gel showed 2 time of swelling ratio compared to the conventional gel at pH of 3, the acidity upon supercritical CO2 contact with water. Besides, a superior rheological property of CO2-responsive gel was observed in the rheology study. Thus CO2-reponsive gel has been considered a potential plugging agent for CO2 flooding.

Understanding the Plugging Performance of HPAM-Cr(III) Polymer Gel for CO2 Conformance Control

Authors:
Sun, X., Missouri University of Science and Technology
Bai, B., Missouri University of Science and Technology

With the demand for conformance control in CO_2 flooding fields, the traditional HPAM-Cr(III) polymer gel has been applied to reduce excess CO_2 production. However, the field application results are mixed. In order to understand the HPAM-Cr(III) polymer gel plugging performance in reservoirs, a numerical reservoir analysis is used to interpret the laboratory results and analyze how the laboratory results are connected to the field application results.

The HPAM-Cr(III) polymer gel plugging performance was evaluated in the laboratory with sandstone core plugs. CO_2 and water effective permeability was measured. Gelant was injected into the cores until the injection pressure stabilized. Breakthrough pressure was tested and followed by two WAG cycles injection. The injection pressure was monitored. Different permeability cores (100- 1200 md) were used to mimic different reservoir conditions. Residual resistance factor was calculated using the stabilized water/CO_2 injection pressure before and after gel treatment.

HPAM-Cr(III) polymer gel has higher breakthrough pressure in the low permeability cores. The polymer gel behavior is discussed using two scenarios: no breakthrough and with breakthrough. If no breakthrough, the numerical reservoir analysis reveals that the polymer gel can successfully divert the CO_2 and water flow to the low permeability layers before breakthrough. The damage to polymer gel is caused by CO_2 diffusion from the low permeability layers. The gel damage rate depends on the diffusion rate. This is the reason that polymer gel treatment can last for a longer time in some fields, In another scenario, CO_2 breakthrough the gel treated formation, CO_2 will penetrate to the depth at where the reservoir pressure gradient in low permeability layer is higher than injection pressure gradient. After this point, CO_2 will be diverted to low permeability layers. In this case, polymer gel has a larger contact area with CO_2 and the gel damage rate will be faster.

Well Integrity Analysis of Injection Wells during CO2 Sequestration, Tarim Basin, China

Authors:
Zhang, J., CNPC Engineering Technology R&D Company Limited
Mu, L., China University of Petroleum
Cai, Z., China University of Petroleum
Guo, Q., CNPC Engineering Technology R&D Company Limited
Tarim Basin is the biggest oil and gas bearing basin in China. The ultra-deep formations in the North Border of Tarim Basin is characterized by ultra-high temperature, high pressure, and ultra-thick composite salt-gypsum formations, are widely distributed, so it will bring great challenge to well integrity during CO_2sequestration. Therefore it is necessary to evaluate the impact of temperature decrease and salt rock creep on injection well integrity.

A numerical model is used to analyze the impact of temperature decrease and salt rock creep on injection well integrity in this process of CO_2 injection and obtain a critical temperature drop value .A set of new ideas for casing selection was proposed based upon long-term corrosion rate of different materials calculated by De. Warrd corrosion prediction model. Well B in Tarim Oilfield was designed in this method, and design results were verified by performing laboratory experiments.

The results show that under some circumstances the micro-annulus between casing and cement sheath can be changed by salt creep or self-healed. The mechanism of casing damage in salt layers was preliminarily revealed. A set of methods is proposed for casing material selection in CO_2 corrosion environment is proposed: corrosion rate of different materials is calculated and converted into long-term corrosion rate. Then the maximum allowable wear extent of the casing is calculated according to casing strength requirement. Finally, casing selection material is determined. The mechanism of casing damage in salt layers was preliminarily revealed during CO_2 sequestration. It is indicated that casing damage is mainly caused by the non-uniform load induced by the creep of salt-gypsum layers ,the low temperature ,and the poor cementing quality.

This new method can effectively reduce cost of casing material selection and make for guaranteeing oilfield development and CO_2 sequestration while maintaining wellbore integrity and controllable well control safety.

Why and How Subsurface Barrier Materials Fail Under Geomechanical Load

Authors:
Du, H., LSU

The carbon capture and storage (CCS) is by far the only technology which can currently reduce the presence of CO_2in the atmosphere on a significant scale. CO_2 storage in depleted oil and gas reservoirs is also not likely to lead to technological difficulties. However, the CCS projects have been limited to a few industrial field-trial types of project, while the large scale industrial deployment is still in the future, [Global CCS Institute, 2015&2019]. One of the major concerns is potential gas leakage to water aquifers or atmosphere due to caprock integrity and leaky wellbores, or simply - failure in barrier materials.

By definition, barrier materials are low permeability rocks or engineered materials, designed to have petrophysical properties that prevent migration of fluids through them. In this respect, shale rock and wellbore cement qualify as barriers at ambient conditions. Once the subsurface pressurization occurs due to CO2 injection, and physical and chemical subsurface equilibrium is out of balance, barrier materials can fail.

This paper presents results from a comparative study of micro/nano geomechanical properties of shale rocks and wellbore cement. Atomic Force Microscopy, Indentation, Scanning Electron Microscopy and Energy Dispersive Spectroscopy were used to provide an insight into susceptibility of barrier materials to fracture initiation/propagation.

Based on the data it can be concluded that geochemistry does influence geomechanical response of barrier materials. In addition, failure at atomic scale is a good indicator for behavior of barrier materials at macro scale, and can be used to predict potential leakage in the field. However, most importantly, by understanding the mechanisms and properties that influence fracture mechanics of natureâ€™s barrier materials, such as shale rocks, we can design and engineer more robust barriers for wellbore construction and abandonment.